INVERSE AND ILL-POSED PROBLEMS SERIES

Monte Carlo Method for Solving
Inverse Problems of Radiation Transfer

Also available in the Inverse and Ill-Posed Problems Series:

Introduction to the Theory of Inverse Problems
A.L. Bukhgeim

Identification Problems of Wave Phenomena - Theory and Numerics
S.I. Kabanikhin and A. Lorenzi

Inverse Problems of Electromagnetic Geophysical Fields
P.S. Martyshko

Composite Type Equations and Inverse Problems
A.I. Kozhanov

Inverse Problems of Vibrational Spectroscopy
A.G. Yagola, I.V. Kochikov, G.M. Kuramshina and Yu.A. Pentin

Elements of the Theory of Inverse Problems
A.M. Denisov

Volterra Equations and Inverse Problems
A.L. Bughgeim

Small Parameter Method in Multidimensional Inverse Problems
A.S. Barashkov

Regularization, Uniqueness and Existence of Volterra Equations of the First Kind
A. Asanov

Methods for Solution of Nonlinear Operator Equations
V.P. Tanana

Inverse and Ill-Posed Sources Problems
Yu.E. Anikonov, B.A. Bubnov and G.N. Erokhin

Methods for Solving Operator Equations
V.P. Tanana

Nonclassical and Inverse Problems for Pseudoparabolic Equations
A. Asanov and E.R. Atamanov

Formulas in Inverse and Ill-Posed Problems
Yu.E. Anikonov

Inverse Logarithmic Potential Problem
V.G. Cherednichenko

Multidimensional Inverse and Ill-Posed Problems for Differential Equations
Yu.E. Anikonov

Ill-Posed Problems with A Priori Information
V.V. Vasin and A.L. Ageev

Integral Geometry of Tensor Fields
V.A. Sharafutdinov

Inverse Problems for Maxwell's Equations
V.G. Romanov and S.I. Kabanikhin

INVERSE AND ILL-POSED PROBLEMS SERIES

Monte Carlo Method for Solving Inverse Problems of Radiation Transfer

V.S. Antyufeev

///VSP///
UTRECHT • BOSTON • KÖLN • TOKYO
2000

VSP BV
P.O. Box 346
3700 AH Zeist
The Netherlands

Tel: +31 30 692 5790
Fax: +31 30 693 2081
vsppub@compuserve.com
www.vsppub.com

First published in 2000

ISBN 90-6764-320-3

Printed in The Netherlands by Ridderprint bv, Ridderkerk.

Contents

Preface

In this book we consider some inverse problems in radiation transfer. Radiation field can be described by the transport equation. The problem is to obtain the equation parameters from the data of radiation transmittance and reflectance. In fact, these parameters are optical parameters of the media.

Particularly, we consider the problems in 1) atmosphere optics; 2) radiation propagation in a plant canopy; 3) tomography in scattering media.

While solving these inverse problems we have to perform direct computation of the radiation field. The prime difficulty is computing the multiscattered radiation intensity. The best way is to use the Monte Carlo method, which is well suited to such computations.

The Monte Carlo method often takes much computer time to obtain good accuracy in computation. So a need arises for effective Monte Carlo developments to compute the field of scattered radiation.

To increase the accuracy of the parameter identification a regularization procedure may be used. But the standard procedure of smooth regularization cannot be applied in this case because the functions to be retrieved are not smooth. For this reason new regularization procedure has been developed.

To solve different inverse problems we also need proper observation data. Sometimes, finding a special observation condition is an essential part of the method for solving the inverse problem.

The plan of the book is as follows.

In the introduction to the first chapter we present elements of the Monte Carlo theory. Then we consider some effective Monte Carlo developments for computation of the scattered radiation field: the method of partial mathematical expectation for slabs and an improvement of this method; estimators that use approximate importance function in modelling free path lengths and direction vectors; optimization in spherical model computation.

In the second chapter a "nonstandard" transport equation is studied. Radiation transfer in stochastic media can be described by a deterministic integral equation of the second kind. Highly peaked phase function causes large error in Monte Carlo estimations of the radiation field. Proper consideration of the process of modelling allows to obtain new probability density functions for the light propagation modelling and a new estimator. This estimator eliminates the large error in computation and speeds up the computation. The results have been applied to computation of the radiation propagation in broken clouds.

In the third chapter we consider radiation field in stochastic media with small fluctuation. A method for calculation of the correlation moments is proposed. To obtain the needed formula we use linear approximation for the stochastic functional. The final formula includes derivatives of the functional over the media parameters. A method for the Monte Carlo computation of these derivatives is described. A theorem is proved stating that the variance is finite.

In the fourth chapter we describe an iterative method for determining of the scattering phase function. Numerical results are presented. The question of theoretical convergence of the proposed method is the subject of our consideration in the second part of the chapter.

Chapter 5 is devoted to regularization methods in solving inverse problems in atmosphere optics. But the standard procedure of smooth regularization procedure cannot be applied in this case because the functions to be retrieved are not smooth. For this reason new regularization procedures has been developed. A special emphasis has been put on inverse problems of determination of the scattering coefficient and phase function.

In Chapter 6 the Monte Carlo method has been applied to the problem of radiation transfer in plant canopy. Balance equations in integro-differential and integral form are derived, and modelling procedure is described in detail. In the second part of the chapter we study inverse problems for the plant model parameters identification. A special method and special observation points have been proposed to retrieve leaf-normal distribution.

In Chapter 7 we consider the problem of computer tomography of media with high scattering level. A mathematical approach has been developed for the investigation of the radiation field. Based on these results, a special source-detector configuration is proposed. It suggests the separation of the intensity of non-scattered and single-scattered radiation from the intensity of the total flux. The conventional Radon transform procedure can be applied to the data processed that way to reconstruct the extinction coefficient distribution. Then the value of the scattering coefficient at any given point can be found separately by solving a system of simple equations. Finally, we present some numerical results to demonstrate a numerical implementation of the method and the distortion of the body image caused by scattering interference.

In Appendix A we study a problem of solving an integral equation of the second kind with unit spectral radius. Such a problem arises in solving some differential equations.

Problems of radiation propagation in optically thick media are of special interest. A new approach is proposed in Appendix B for solving this problem by the Monte Carlo method.

The method for identification of the phase function in Chapter 4 gives good results for comparatively small values of optical thickness of the atmosphere. A new iterative method has been developed recently for solution of this important problem of the atmosphere optics. We discuss this method in Appendix C.

Chapter 1.

Monte Carlo modifications for solving problems of radiation transfer

In the first section we cite in brief some results from the theory of the Monte Carlo method we will use in further considerations.

Particularly, we present the following topics:

1) the Monte Carlo estimation of linear functionals of solution of the transport integral equation of the second kind;
2) the usage of the conjugate Monte Carlo scheme;
3) weight methods and the importance function.

The details can be found in Spanier and Gelbard (1969), Albert (1956), Ermakov (1975), Marchuk, *et. al.* (1980), Sobol (1973), Mikhailov (1974, 1992).

1.1. ELEMENTS OF USING THE MONTE CARLO METHOD IN RADIATION TRANSFER COMPUTATION

To find characteristics of the scattered radiation field by the the Monte Carlo method (method of statistical modelling) the transport equation should be represented in the form of integral equation of the second kind

$$f(x) = \int_X k(x', x) f(x') dx' + h(x) \tag{1.1}$$

with the kernel

$$k(x', x) = \frac{\sigma_s(r')\sigma(r)g(\omega', \omega)\exp(-\tau(r', r))}{2\pi\sigma(r')|r - r'|^2}\delta_s(\omega) \tag{1.2}$$

and the source $h(x)$.

This is an equation for the collision density $f(x)$ of the particles in the phase space. The collision density is connected with the particle flux density $\Phi(r, \omega)$ (which is a solution of the integro-differential transport equation) by the simple relation

$$f(r, \omega) = \sigma(r)\Phi(r, \omega).$$

Here we use the following notation:

$X = R \times \Omega$ is the phase space of the equation;

$x = (r, \omega)$ are generic points of the phase space;

R is the geometric space of collision points $r = (x, y, z)$ with the Cartesian coordinates;

Ω is the space of unit vectors $\omega = (a, b, c)$ of the direction of particles immediately before their collision;

$s(r', r)$ is the direction vector from r' to r:

$$s(r', r) = \frac{r - r'}{|r - r'|};$$

$\sigma_s(r)$ is the scattering cross-section coefficient;

$\sigma_a(r)$ is the absorption cross-section coefficient;

$\sigma(r)$ is the total extinction coefficient:

$$\sigma(r) = \sigma_s(r) + \sigma_a(r);$$

λ is the probability of the photon survival at a collision at the point r:

$$\lambda = \frac{\sigma_s(r)}{\sigma_s(r) + \sigma_a(r)};$$

$\tau(r_1, r_2)$ is the optical distance between the points r_1 and $r_2 = r_1 + \omega d$:

$$\tau = \int_0^d \sigma(r_1 + \omega t)dt.$$

the standard case

$$\sigma(r(l))\exp(-\tau(l))$$

the probability density of the free path length of a photon tween two successive collisions. Here

$$\tau(l) = \int_0^l \sigma(r(t))dt$$

the optical length of the free path.

ι) is the phase function (indicatrix), where $\mu = (\omega', \omega)$, ω' and are particle directions before and after the scattering at the int r, correspondingly. The phase function is the probability nsity function for the direction transition $\omega' \to \omega$. This density normed to unity by the condition

$$\int_{-1}^{1} g(\mu)d\mu = 1.$$

Monte Carlo computation a piecewise approximation for the ase function $g(\mu)$ is used with given values

$$g(\mu_i) = g_i \geq 0, \quad i = 1, \ldots, n.$$

some points (nodcs) μ_i in the interval $[-1, +1]$:

$$1 = \mu_1 > \mu_2 > \ldots > \mu_n = -1$$

ιe phase function is interpolated linearly between the points. ιus, the phase function can be specified by a vector

$$g = (g_1, \ldots, g_n).$$

the "generalized" transport equation (see Chapter 2) the scat- ing phase function is a function of two unit vectors ω' and ω.

c) is the density of points x at the first collisions in the phase

Different characteristics of the scattered radiation field may be often represented as linear functionals of the collision density f:

$$I_\varphi(f) = (f, \varphi) = \int_X f(x)\varphi(x)dx.$$

We only need to find an appropriate "weight" function φ, perhaps, a generalized one.

For instance, to find a radiation flux $\Phi(r^*, \omega^*)$ at a point (r^*, ω^*) in the phase space, where r^* is a point of the detector location and ω^* is a vision direction, we may take

$$\varphi(r, \omega) = \delta_{r^*}(r) \cdot \delta_{\omega^*}(\omega).$$

Further, we will assume without mentioning that the conditions are fulfilled for convergence of the Neumann series that represents the solution of equation (1):

$$f = h + Kh + K^2h + \dots .$$

Here K is an integral operator with the kernel (2).

We may compute linear functionals of $I_\varphi(f)$ by the the Monte Carlo method modelling photon trajectories. If $\{x_n\}$ is a photon trajectory (i.e. a Markov chain of photon collisions in the phase space) then the value I_φ may be represented in the form of mathematical expectation

$$I_\varphi = M_\xi,$$

where

$$\xi = \varphi(x_0) + \varphi(x_1) + \varphi(x_2) + \dots + \varphi(x_n). \tag{1.3}$$

Here x_0 is a random point in the region of the physical source. It should be modelled according to the source density function $h(x_0)$. x_n is a random point of the trajectory termination. The particle either absorbs at this point or goes out of the region after a collision at this point. Transition from the point x_i to x_{i+1} should be modelled according to the density function $k(x' \to x) \equiv k(x', x)$; $\varphi(x)$ is a "contribution function" calculated at the corresponding collision.

We have noticed above that the "contribution function" $\varphi(x)$ under the integral may be a generalized one. This is the case where the physical detector is located in a geometrical point or where it has a single vision direction. Such statistical estimators are usually called "local" estimators. It is clear that we cannot calculate any value of the function $\varphi(x)$. For this case special modifications of the standard computation algorithms should be developed, which leads to changes in trajectory modelling and to a regular contribution function $\varphi(x)$.

1.2. THE "CONJUGATE" COMPUTATION SCHEME

Sometimes it is more convenient to model trajectories using the conjugate equation

$$f(x) = \int_X k(x, x')f(x')dx' + \varphi(x) \tag{1.1'}$$

instead of the equation (1). This is the so-called "conjugate scheme" of the Monte Carlo method. In the conjugate scheme we should start photon trajectory in the region of a physical radiation detector. Then we model the photon transitions $x' \to x$ in the phase space and acts of absorption the same way as in the case of use of the standard transport equation, since the kernel of the transfer equation is symmetric. The contribution function also changes. It is proportional to the source density function

$$\varphi' = \text{const} \cdot h.$$

Here we give a few examples of possible advantages in the use of the conjugate scheme instead of the direct one.

1. Consider the problem of calculating the radiation field in the Earth's atmosphere. In this case the statistical estimate of the contribution of collisions in the "direct scheme" includes the factor (the attenuation coefficient), which changes essentially over the altitude. This results in large statistical errors. The corresponding contribution function in the conjugate scheme does not include such a factor. Therefore, using the conjugate scheme in this case we obtain more accurate statistical estimate than in the "direct scheme" at the same number of modelled trajectories.

2. As it was said above, we cannot use the common Monte Carlo scheme if the radiation source is concentrated in a point since the contribution function is the generalized delta-function in this case. However, if the radiation source is distributed over the space (i.e. h is a regular function), the use of the conjugate scheme allows to easily get around this obstacle.

3. In the case of symmetry of the system "medium-source" we can use the same trajectories to perform simultaneous computation of different desired functionals.

For details see Marchuk, *et. al.* (1980), Antyufeev and Mikhailov (1976), Antyufeev, Ivanov, *et. al.* (1980).

1.3. USE OF THE IMPORTANCE FUNCTION

When we model trajectories according to the densities $\widetilde{h}(x)$ and $\widetilde{k}(x',x)$ different from the physical densities $h(x)$ and $k(x',x)$, respectively, then the estimate (3) takes the form

$$\xi' = Q_0\varphi(x_0) + Q_1\varphi(x_1) + \ldots + Q_n\varphi(x_n), \tag{1.4}$$

where

$$Q_0 = \frac{h(x_0)}{\widetilde{h}(x_0)}, \quad Q_n = Q_{n-1}\frac{k(x_{n-1},x_n)}{\widetilde{k}(x_{n-1},x_n)}. \tag{1.5}$$

Let the "importance" function f^*, which is a solution to the equation conjugate to (1):

$$f^*(x) = \int_X k(x,x')f^*(x')dx' + \varphi(x)$$

be approximately known. Take the density functions $h(x)$ and $k(x',x)$ close to the "ideal" densities

$$h^*(x) = \frac{h(x)f^*(x)}{(h,f^*)} \tag{1.6}$$

and

$$k^*(x',x) = \frac{k(x',x)f^*(x)}{[K^*f^*](x')}, \tag{1.6'}$$

respectively. Then, modelling the chain of collisions according to $h(x)$ and $k(x',x)$ we arrive at an estimate of the form (5) with the variance close to zero. However, we should note that it is not only difficult to find the densities $h(x)$ and $k(x',x)$ but to model them as well.

Note that we can find $f^*(x)$ as the value of the functional $I_\varphi(f_x)$, where f_x is a solution to the equation

$$f_x(y) = \int_X k(y',y)f_x(y')dy' + \delta_x(y)$$

with the source δ_x concentrated at the point x of the phase space. This follows directly from the relation

$$(f,\varphi) = (f^*,h),$$

since

$$f^*(x) \equiv (f^*, \delta_x) = (f_x, \varphi) \equiv I_\varphi(f_x).$$

This approach implies the following. If a particle suffers a collision at the point x of the phase space, then we should calculate the average contribution to the estimate from all the trajectories starting at the point x. In fact, this contribution is just equal to $f^*(x)$. Then we should model the next collision according to the modified density (6′). Finally, we should multiply the contribution of the collision by the corresponding weight factor of the form (5).

Hence, it follows that in modelling the trajectories using the direct equation with a point source at the point x of the phase space we may in fact find the solution $f^*(x)$ of the adjoint equation at this point. Therefore, using the conjugate scheme for solving the problem and modelling the trajectories starting at the point x, we calculate the solution $f(x)$ to the direct equation. Hence, unlike the "direct scheme", to optimize calculations we can use here one value of the importance function in calculating several functionals.

In the next section we discuss how the approximate importance function can be used in calculating the radiation transfer in plane and spherical layers of small optical thickness.

To solve the inverse problems which are considered in the following chapters, we need first to solve "direct" problems, that is, to compute angular distribution of the scattered solar radiation at given points of the phase space by the Monte Carlo method. This is the most time-consuming part of algorithm for solving an inverse problem. It takes a lot of computer time to obtain appropriate accuracy.

In the next section we study some improvements of the standard Monte Carlo scheme to optimize the "direct" computation.

Note that we compute by the Monte Carlo method a decomposition of radiation intensity according to the number of scattering acts. In this sum the intensity of single-scattered radiation may be represented as a one-dimensional integral that is easy to calculate by a simple quadrature formula or, maybe, even analytically. So, the proposed developments are concerned with optimization for computing the multiscattered radiation intensity.

1.4. METHOD OF MATHEMATICAL EXPECTATION FOR PLANE SLAB MODEL

In practical computation the spherical Earth's atmosphere may be changed for a plane slab. That is why consideration of a slab geometry is so important.

Consider a problem of computing the solar scattered radiation intensity in a homogeneous plane slab. The monodirectional source is located on the upper boundary plane. For simplicity assume the following.

1) Albedo (reflection coefficient) of the bottom plane is equal to zero.

2) There is no absorption.

3) The phase function is independent of the altitude z.

The first and second restrictions may be easily removed in practical computation. The third restriction is more significant. To remove it, we may consider a phase function step over the altitude z.

So, we consider a plane layer $\{(x, y, z) : \ 0 \le z \le 1\}$. The problem is to estimate by the Monte Carlo method a linear functional

$$I = (\Phi, F)$$

of the radiation flux $\Phi(x)$, $\Phi(x) = f(x)/\sigma(r)$, where $\sigma(r)$ is a scattering coefficient, $f(x)$ is a collision density function that satisfies the integral transport equation

$$f(x) = \int\limits_{X'} f(x')k(x' \to x)dx' + \psi(x), \tag{1.7}$$

the scattering coefficient is

$$\sigma(r) = \begin{cases} \sigma(z), & 0 \le z \le t \\ 0, & \text{otherwise}, \end{cases}$$

$$\tau(r', r) = \int\limits_{z'}^{z} \frac{\sigma(t)}{c} dt,$$

ψ is a density function for first-order collisions originated by a physical source

$$\psi(r, \omega) = \xi(\omega) \frac{\sigma(z)}{c} \exp\left[-\int\limits_{z}^{t} \frac{\sigma(t)}{c} dt \right] \delta_u(\omega),$$

where the direction vector u is

$$u = \frac{r - e_z}{|r - e_z|},$$

$\xi(\omega)$ is the density function for the initial particle directions.

It is known that at a given optical thickness of the layer

$$\tau = \int_0^t \sigma(z)dz$$

the solution to that problem is independent of the change of $\sigma(z)$ over the altitude z, therefore we take for simplicity

$$\sigma(z) = \begin{cases} \sigma, & 0 \le z \le t \\ 0, & \text{otherwise.} \end{cases}$$

The value I may be rewritten in the form

$$I = (f, \varphi),$$

where φ is a weight function. Note that in horizontally-homogeneous problem φ is independent of x and y:

$$\varphi(r, \omega) = \varphi(z, \omega).$$

Denote for brevity

$$\int_{-\infty}^{\infty}\int_{-\infty}^{\infty} f(x, y, z, \omega)dx\, dy = p(z, \omega),$$

$$\int_{-\infty}^{\infty}\int_{-\infty}^{\infty} \psi(x, y, z, \omega)dx\, dy = \theta(z, \omega).$$

Then the value I can be represented in the form

$$I = \int_{\Omega}\int_{-\infty}^{\infty}\int_{-\infty}^{\infty}\int_{-\infty}^{\infty} f(x, y, z, \omega)\varphi(z, \omega)dz\, d\omega$$

$$= \int_{\Omega}\int_{-\infty}^{\infty} p(z, \omega)\varphi(z, \omega)dz\, d\omega = (p, \varphi).$$

Transform the integral in the right-hand side of equation (7). Change the Cartesian system with the variable $r = (x, y, z)$ for the polar system with a pole at the point r and the variables

$$\lambda = |r' - r| > 0$$

and

$$s = \frac{(r' - r)}{|r' - r|} \in S,$$

where S is a unit sphere. Then we obtain

$$\int\limits_{X'} f(x')k(x' \to x)dx'$$

$$= \int\limits_{\Omega}\int\limits_{R} f(r',\omega')\frac{\sigma(z)\exp(-\tau)}{|r-r'|^2} \cdot \frac{g(\omega\cdot\omega')}{2\pi} \cdot \delta_s(\omega)d\omega' dr'$$

$$= \int\limits_{\Omega} \frac{g(\omega\cdot\omega')}{2\pi} \int\limits_{0}^{\infty}\int\limits_{S} d\lambda\, ds\, f(r+\lambda s,\omega)\sigma(z)\exp(-\tau)\delta_s(\omega)$$

$$= \int\limits_{\Omega} \frac{g(\omega\cdot\omega')}{2\pi} \int\limits_{0}^{\infty} \lambda\, f(r-\lambda\omega,\omega')\sigma(z)\exp\left[-\int\limits_{0}^{\lambda}\sigma(z-tc)dt\right].$$

Replace the integral in equation (7) with the expression just received. Integrating this equation over x and y we obtain the equation for $p(z,\omega)$:

$$p(z,\omega) = \int\limits_{\Omega} \frac{g(\omega\cdot\omega')}{2\pi} \int\limits_{0}^{\infty} \lambda\, p(r-\lambda\omega,\omega')\sigma(z)\exp\left[-\int\limits_{0}^{\lambda}\sigma(z-tc)dt\right] + \theta(z,\omega).$$

Change the variables

$$\lambda \to z' = z - \lambda c.$$

Then

$$p(z,\omega) = \int\limits_{\Omega} \frac{g(\omega\cdot\omega')}{2\pi} \int\limits_{\varepsilon}^{z} dz'\, p(z',\omega')\frac{\sigma(z)}{|c|}\exp\left[-\int \frac{\sigma(t)}{c}dt\right] + \theta(z,\omega).$$

Further we assume that $0 \le z \le 1$ and, hence, $\sigma(z) = \sigma$ (if $z \notin [0,1]$ then $p(z,\omega) = 0$). Therefore, the equation takes the form:

$$p(z,\omega) = \int\limits_{\Omega} \frac{g(\omega\cdot\omega')}{2\pi} \int\limits_{\varepsilon}^{z} dz'\, p(z',\omega')\frac{\sigma}{|c|}\exp\left[-\frac{\sigma|z'-z|}{|c|}dt\right] + \theta(z,\omega).$$

Here the limits of integration are as follows:

$$\begin{cases} c>0: & \int\limits_0^z dz', \\ c<0: & \int\limits_z^1 dz'. \end{cases}$$

Denote for brevity

$$l(z,\omega) = \begin{cases} \frac{\sigma}{c}\exp\left(-\frac{\sigma z}{c}\right), & 0 \le z \le 1, \\ 0, & \text{otherwise.} \end{cases}$$

Finally, we obtain

$$p(z,\omega) = \int\limits_{\Omega} \frac{g(\omega\cdot\omega')}{2\pi} \int dz'\, p(z',\omega') l(z-z',\omega') + \theta(z,\omega),$$

or, in the operator form,

$$p = Lp + \theta,$$

where L is an integral operator and

$$\theta(z,\omega) = \xi(\omega)\frac{\sigma}{c}\exp\left[\frac{-\sigma(1-z)}{|c|}\right],$$

$0 \le z \le 1$, $c < 0$. A solution to that integral equation may be represented as a Neumann series:

$$p = \sum_{n=0}^{\infty} L^n\theta.$$

Rewrite the n-th term of that series in detail:

$$L^n\theta(z,\omega) = \int\limits_{\Omega}\dots\int\limits_{\Omega} d\omega_0\dots d\omega_{n-1}\xi(\omega_0)\int\dots\int dz_0\dots dz_{n-1}$$
$$\frac{g(\omega_0\cdot\omega_1)\dots g(\omega_{n-1}\cdot\omega)}{(2\pi)^n}\frac{\sigma}{|c_0|}\exp\left[-\frac{\sigma(1-z_0)}{|c_0|}\right]$$
$$\times\, l(z_1-z_0,\omega_1)\dots l(z_{n-1}-z_{n-2},\omega_{n-1})l(z-z_{n-1},\omega).$$

Taking into account that the phase function $g(\mu)$ is independent of the spatial variable z, we can carry the factor $g(\omega_0 \cdot \omega_1) \dots g(\omega_{n-1} \cdot \omega)$ out from the integral with respect to z:

$$\dots = \int_\Omega \dots \int_\Omega d\omega_0 \dots \omega_{n-1} \xi(\omega_0) \frac{g(\omega_0 \cdot \omega_1) \dots g(\omega_{n-1} \cdot \omega)}{(2\pi)^n}$$

$$\times \int \dots \int dz_0 \dots dz_{n-1} \frac{\sigma}{|c_0|} \exp\left[-\frac{\sigma(1-z_0)}{|c_0|}\right]$$

$$\times\, l(z_1 - z_0, \omega_1) \dots l(z_{n-1} - z_{n-2}, \omega_{n-1}) l(z - z_{n-1}, \omega).$$

Denote for brevity

$$q_n(z, \omega_0, \dots \omega_{n-1}, \omega)$$

$$= \int \dots \int dz_0 \dots dz_{n-1} \frac{\sigma}{|c_0|} \exp\left[-\frac{\sigma(1-z_0)}{|c_0|}\right]$$

$$\times\, l(z_1 - z_0, \omega_1) \dots l(z_{n-1} - z_{n-2}, \omega_{n-1}) l(z - z_{n-1}, \omega).$$

Here the value of the integral $q_n(z, \omega_0, \dots \omega_{n-1}, \omega)$, in fact, depends only on the variables $c_0, c_1, \dots, c_{n-1}, c$, which are coordinates of the directions $\omega_0, \omega_1, \dots, \omega_n, \omega$:

$$q_n(z, \omega_0, \dots \omega_{n-1}, \omega) \equiv q_n(z, c_0, c_1, \dots c_{n-1}, c).$$

The proposed development is based on the following remark. This integral can be calculated analytically at fixed $c_0, c_1, \dots c_{n-1}, c$, not by the numerical Monte Carlo method, which leads to significant reduction in computation error.

It is easy to understand that the required integral is a function of the following form

$$q_n(z, c_0, \dots c_{n-1}, c) = \sum A_i^n \exp(-\alpha_i z).$$

We will prove it by induction and find recurrent formulas for the coefficients A_i^n.

For $n = 0$ we have:

$$q_0(z, c) = \frac{\sigma}{|c|} \exp\left(\frac{\sigma(1-z)}{|c|}\right) = A_0^0 \exp(-\alpha_0 z).$$

Here

$$\alpha_0 = \frac{\sigma}{c}, \qquad A_0^0 = -\alpha_0 \exp(-\alpha_0).$$

Assume that for $n = k$

$$q_{n-1}(z, c_0, \dots c_{n-2}, c) = \sum_{i=0}^{n-1} A_i^{n-1} \exp(-\alpha_i z),$$

where the coefficients α are

$$\alpha_i = \frac{\sigma}{c_i}, \quad i = 0, 1, \dots, n-2, \qquad \alpha_{n-1} = \frac{\sigma}{c}.$$

Then

$$\begin{aligned} q_n(z, c_0, \dots c_{n-1}, c) &= \int_{\varepsilon}^{z} q_{n-1}(z', c_0, \dots, c_{n-2}, c)\, l(z - z', \omega) dz' \\ &= \sum_{i=0}^{n} A_i^n \exp(-\alpha_i z). \end{aligned}$$

Calculate recurrently the coefficients A_i^n, $i = 0, 1, \dots, n-1$ using A_i^{n-1}. At $c > 0$ we have:

$$A_i^n = A_i^{n-1} \frac{\alpha_n}{\alpha_n - \alpha_i}, \quad i = 0, 1, \dots, n-1, \tag{1.8}$$

$$A_n^n = \sum_{i=0}^{n-1} A_i^{n-1} \frac{\alpha_n}{\alpha_n - \alpha_i}, \quad i = 0, 1, \dots, n-1.$$

At $c < 0$ we have:

$$A_i^n = -A_i^{n-1} \frac{\alpha_n}{\alpha_n - \alpha_i}, \quad i = 0, 1, \dots, n-1,$$

$$A_n^n = \sum_{i=0}^{n-1} A_i^{n-1} \frac{\alpha_n}{\alpha_n - \alpha_i}, \quad i = 0, 1, \dots, n-1.$$

Thus, for a fixed chain of flight directions of the particle trajectory $\omega_0, \dots, \omega_{n-1}, \omega$ the density of n-th order collisions over the altitude z can be calculated precisely, which significantly reduces the variance of the Monte Carlo estimate. We also need not model the particle path length from one collision to another.

Thus, in the new method we do not model free path length and geometrical coordinates as well. But the "trajectories" still exist. The trajectory is a chain of the consequent directions $\omega_0, \omega_1, \dots$. Of course we cannot terminate this trajectory in "natural" way after the particle is absorbed or leaves the region (layer). That is why we should terminate the trajectory in "artificial" way. Below we consider two ways of doing that.

1. Termination "without leaving".
The density function for the scattered particles at a collision is calculated according to the given formulas. We terminate the trajectory in n-th step where n is chosen in the following way. Take a "threshold" $\varepsilon > 0$ and find such n that $\|L^n\| < \varepsilon$.

Another way is as follows. At every step we calculate a "weight" of the distribution, that is, the integral

$$W_n = \int_0^T q_n(z, \omega_1, \dots, \omega_n) dz > 0.$$

We terminate the trajectory at the moment the weight W_n becomes less than the fixed "threshold" ε.

2. Termination "with leaving".
This is a randomized termination of the trajectory. At every collision we calculate the weight W_n of the distribution. This value determines the share of the particles that do not leave the layer and scatter inside it. Model a random number α uniformly distributed on $[0, 1]$. If $W_n < \alpha$ then we terminate the trajectory. In the case $W_n > \alpha$ we act the following way. We should divide the coefficients A_i^n in the expression for the distribution q_n by the weight W_n, so that the distribution "weight" remains equal to unity. Then we resume modelling the trajectory.

If there exists a bottom plane we can model in the same way the type of scattering that is "scattering in the atmosphere" or "scattering on the bottom surface".

A few remarks about the proposed method are given below.

1. This method may be called a method of "integration forward" over the whole trajectory. However, we may limit integration to a fixed number of steps forward. If the optical thickness of the layer is small, we may even perform just one integration forward.

The matter is that the proposed method takes much more computer time to model a trajectory compared with the standard one. We take advantage in computing the multiscattered radiation intensity. But in an optically thin layer this value is small, so one to two integrations forward will be enough.

As the thickness of the layer grows, the efficiency of the "total" integration should increase. However, numerical experiments have shown that the efficiency decreases at large optical thickness. The cause of this has been studied, which resulted in the improvement of the method. The results of the study and the improved method are described in the following section.

3. The proposed method allows to calculate the collision density of photons over the altitude, and, consequently, different linear functionals of that density (not only intensity).

4. The obtained expression for a collision density is a precise one in case of a plane layer. But this expression can be used in computation for a spherical model for approximating the importance function (see Section 1.7).

Numerical experiments have shown that the use of the proposed method increases efficiency from 4 to 5 times.

1.5. IMPROVEMENT OF THE METHOD OF MATHEMATICAL EXPECTATION

In the previous section we have considered an efficient Monte Carlo development for calculating the radiation transfer in a plane layer. We have found the distribution function $q_n(z)$ for the n-th collision point over the altitude z, provided the directions in the previous part of the trajectory are fixed. This function is given by the formula

$$q_n(z, \omega_1, \dots \omega_n) = \sum_{i=1}^{n} A_i^n \exp\left(\frac{-z\tau}{c_i}\right), \tag{1.9}$$

where τ is the optical thickness of the layer, ω_i are the direction vectors of the previous part of the trajectory ($\omega_i = (a_i, b_i, c_i)$), and the coefficients A_i^n ($i = 1, 2, \dots, n$) are calculated recurrently using A_i^{n-1} ($i = 1, 2, \dots, n-1$). Thus, the density $q_n(z)$ can be represented as a linear combination of the "basic" functions $\exp(-zt/c_i)$, ($i = 1, \dots, n$). Hence, it is easy to obtain a formula for the contribution function at each collision.

The method is quite efficient for small τ. However, numerical experiments have shown that its efficiency diminishes as τ grows. At $\tau > 10$ the computer program terminates because of overflow. Investigation has shown that the absolute values of the coefficients A_i^n grow dramatically as τ grows. Although in adding the terms in (9) can be cancelled, this addition causes large errors in computation. We may perform the computation with double accuracy, which delays the error but does not eradicate it.

To resolve the problem, we should consider in greater detail the algorithm for recurrent calculation of the coefficients A_i^n.

Now assume that the density q_n is given by the formula (9) and ω_{n+1} is the direction of the next free path. Consider the case $c_{n+1} > 0$, namely

$$
\begin{aligned}
&q_{n+1}(z, \omega_1, \dots \omega_n) \\
&\quad = \int_0^z dz' \frac{\tau}{c_{n+1}} \exp\left(\frac{(z'-z)\tau}{c_{n+1}}\right) q_n(z', \omega_1, \dots, \omega_n) \\
&\quad = \sum_{i=1}^n A_i^n \frac{\tau}{c_{n+1}} \int_0^z dz' \exp\left(\frac{(z'-z)\tau}{c_{n+1}}\right) \exp\left(\frac{-z'\tau}{c_i}\right) \\
&\quad = \sum_{i=1}^n A_i^n \frac{\tau}{c_{n+1}} \cdot \frac{\exp\left(\dfrac{-z\tau}{c_i}\right) - \exp\left(\dfrac{-z\tau}{c_{n+1}}\right)}{\dfrac{\tau}{c_{n+1}} - \dfrac{\tau}{c_i}}
\end{aligned}
$$

Decompose each difference quotient into a sum of fractions:

$$
\begin{aligned}
&\frac{\exp\left(\dfrac{-z\tau}{c_i}\right) - \exp\left(\dfrac{-z\tau}{c_{n+1}}\right)}{\dfrac{\tau}{c_{n+1}} - \dfrac{\tau}{c_i}} \\
&\qquad = K_i^{n+1} \exp\left(\frac{-z\tau}{c_i}\right) + (-K_i^{n+1}) \exp\left(\frac{-z\tau}{c_{n+1}}\right). \qquad (1.10)
\end{aligned}
$$

Next we group together the terms including the same exponents and calculate the coefficients A_i^{n+1} of the exponents.

Pay attention to the expression (10). If $|\tau/c_i - \tau/c_{n+1}|$ is close to zero, then the absolute value of the coefficient K_i^{n+1} is large. This coefficient enters into A_i^{n+1} as a term. The accuracy of summing the coefficients K_i^{n+1} is not high because the orders of their magnitudes differ greatly. Thus, for example, when calculating on a computer we have $10^{10} + 1 = 10^{10}$ and, consequently, $(10^{10} + 1) - 10^{10} = 0$. Besides, computation of the coefficient A_i^{n+1} may cause overflow.

To overcome this difficulty, we proceed as follows. We would not group together the terms the way we did it above, but consider the fractions

$$\frac{\exp x - \exp y}{x - y}$$

and note that if $|x - y| \sim 0$ then

$$\frac{\exp x - \exp y}{x - y} \sim \exp x. \tag{1.11}$$

Using this relation, we rewrite (10) as

$$\frac{\exp\left(\dfrac{-z\tau}{c_i}\right) - \exp\left(\dfrac{-z\tau}{c_{n+1}}\right)}{\dfrac{\tau}{c_{n+1}} - \dfrac{\tau}{c_i}} \approx z \exp\left(\frac{-z\tau}{c_i}\right).$$

For the case $c_{n+1} < 0$ we obtain the same results.

Thus, to improve the algorithm, we should extend the set of "basic" functions $\exp(-z\tau/c_i)(i = 1, \dots, n+1)$ by adding the functions $z\exp(-z\tau/c_i)$. Now the density q_n can be represented as

$$q_n(z, \omega_1, \dots \omega_n) = \sum_{i=1}^{n} \left(A_i^n \exp\left(\frac{-z\tau}{c_i}\right) + B_i^n \exp\left(\frac{-z\tau}{c_i}\right)\right)$$

(some coefficients B_i^n may be zero in the expansion). Accordingly, the recurrent formulas for the transition from the set A_i^n, B_i^n $(i = 1, 2, \dots, n)$ to the set A_i^{n+1}, B_i^{n+1} $(i = 1, 2, \dots, n+1)$ also change.

The "threshold"

$$\varepsilon = \left| \frac{\tau}{c_i} - \frac{\tau}{c_j} \right|$$

should be chosen in advance in practical calculations. Numerical experiments have shown that $\varepsilon = 0.5$ provides high accuracy in the calculations. The error level was less than 0.1% and the accuracy did not fall as τ grew.

In the next two sections we consider the modifications of the standard Monte Carlo scheme based on the use of the approximate importance function. In the following section we use the obtained approximation to modify free path length distribution. In the next section we modify distribution of the flight direction.

1.6. MODIFICATION OF THE DISTRIBUTION OF PARTICLE PATH LENGTH

This modification is also based on the results obtained in Antyufeev (1979) and the results considered in the previous section.

First we consider the problem for a plane homogeneous layer. At the end of the section we consider how to reduce a problem for an inhomogeneous spherical layer to the above problem.

Note that in modelling a trajectory the free path length and flight direction are modelled independently. Hence, we can proceed as follows. First, according to physical density, we can model the Markov chain $\omega_1,\ \omega_2,\dots$ of flight directions and then model the chain $l_1,\ l_2,\dots$ of the corresponding length. Besides, we should find an approximate expression for the average contribution of trajectories originating at an arbitrary fixed point r to the statistical estimate.

In the preceding section we have presented analytical formula for calculating the density of the n-th particle collision at the altitude z. The formula allows one to calculate easily the precise average contribution of the first n particle collisions to the statistical estimate of intensity at a fixed initial point $x=(z,\omega)$ and fixed chain $\omega_1,\ \omega_2,\dots$ of the subsequent path directions. The corresponding expression is of the form:

$$\sum_{i=1}^{n} A_i^n \exp\left(\frac{-z}{c_i}\right) + A_{n+1}^n \exp\left(\frac{-z}{c^*}\right). \tag{1.12}$$

For the plane layer these formulas provide a significant speed-up in computation. In doing so, it is sufficient to model only the chain of flight directions without modelling the corresponding lengths. But we can also use the lengths for optimizing the calculations for both plane and spherical layers, namely for appropriate n the formula (12) yields an approximate expression for the "partial importance function" at the point x with high accuracy. Here the term "partial" implies that we deal not with the real importance function but with a "conditional" one, which refers to the trajectories with the fixed sequence of the vectors $\omega_1,\ \omega_2,\dots$.

Naturally, in the case of a plane layer for a fixed chain of directions we can obtain theoretically the precise result with zero variance. However, it is not reasonable to use the formula (9) with a large number of terms because the "modelling according to importance" and calculation of the corresponding weight factors require a great deal of computing time and the advantage of the modification will be reduced to zero. For an optically thin layer, in the sum (9) we can restrict ourselves to the value $n = 1$ and in doing so we obtain fair accuracy.

Consider, for example, the formula obtained for the free path length density and for the weight factor at $c > 0$ and $c^* < 0$. In this case the "physical" density of the path length modelling is equal to

$$\frac{1}{c}\exp\left(\frac{-(t-z)}{c}\right), \quad t \geq z. \tag{1.13}$$

The approximate importance function is equal to the contribution function up to the z-independent factor:

$$\varphi(z, c^*) = \begin{cases} \exp\left(\frac{z}{c^*}\right), & 0 \leq z \leq H, \\ 0, & z > H. \end{cases} \tag{1.14}$$

The product of the functions (13) and (14) should be normed to unity on the segment $[z, H]$. The normed modified density is equal to

$$\frac{1}{c}\exp\left(-t\left(\frac{1}{c}-\frac{1}{c^*}\right)\right).$$

Here the cumbersome formula for the norming constant

$$C = \frac{1}{\frac{1}{c}-\frac{1}{c^*}}\left(\exp\left(\frac{H}{c^*}\right)\exp\left(-\frac{H-x}{c}\right)-\exp\left(\frac{z}{c^*}\right)\right)$$

is due to the fact that we model the path length assuming that the particle does not leave the layer. The weight factor is equal to

$$C\frac{\frac{1}{c}\exp\left(-\frac{t-z}{c}\right)}{\exp\left(-t\left(\frac{1}{c}-\frac{1}{c^*}\right)\right)} = C\frac{1}{c}\exp\left(\frac{z}{c}-\frac{t}{c^*}\right).$$

Similar formulas can be obtained for other signs of the values of the coordinates c and c^* and for $n \leq 2$.

Now consider the problem for a spherical horizontal homogeneous layer. It is well known that a spherical layer may be replaced by the corresponding plane layer without significant loss in accuracy if the optical thickness in the vertical direction is not high.

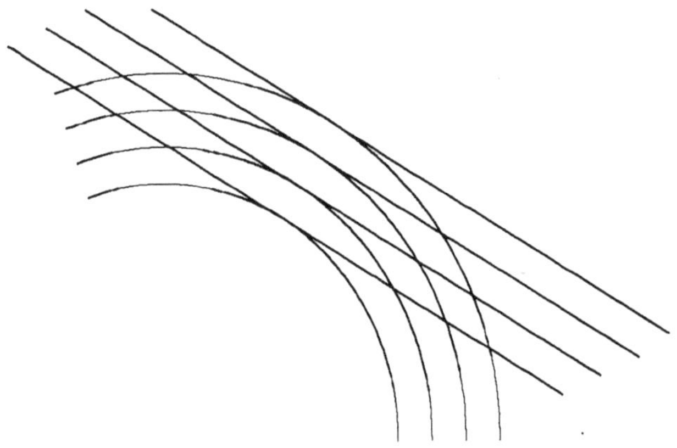

Figure 1: Mutual arrangement of the spherical and plane boundary surfaces

Let the radiation detector be located at a point lying inside or at the boundary surface of the atmosphere. Assume that the Earth's center is the origin and draw two planes tangent to the inner and outer boundary surfaces of the atmosphere and orthogonal to the radius vector of the detector (Figure 1). Then the corresponding layer lies just between these two planes. The considered vertical layer is, generally speaking, vertically inhomogeneous, and the figure presents the corresponding vertical division of the tangent plane layer. However, the plane problem under consideration is one-dimensional with respect to the space variables and therefore we can replace the inhomogeneous layer by the homogeneous one.

Thus, the problem with a vertically inhomogeneous spherical layer may be reduced to that with a homogeneous plane layer with high accuracy. Hence, in this case we can use the constructed approximation to the "partial importance function". It should be remembered that the points of the spherical layer correspond to the points of the tangent plane layer having the same optical (not geometrical) altitude above underlying surface.

1.7. MODIFICATION OF THE TRAJECTORY ANGULAR DISTRIBUTION

Here we consider another approximation to the importance function. The use of this approximation leads to a modification in modelling the direction of the particle trajectory.

"Modelling-according-to-importance" method is a universal one. Theoretically, we can utilize it any time we solve an integral equation of the second kind by the Monte Carlo method. Theoretically, we can reduce variance of any statistical estimate to zero. However, in practical calculation we normally face serious obstacles in using it. The problem of evaluating the importance function in all points of the phase space is often much more difficult than the original one. Naturally, in practical computation we would rather try to find a suitable approximation to the importance function. Such an approximation should satisfy the following conditions.

1. The computation procedure for the importance function should not be very time-consuming, otherwise it would reduce the advantage of its use.

2. The relevant data should have compact form. What we mean is that the phase space in radiation transport problems is multidimensional. So, we cannot record all the numerical data for the importance function because of the lack of computer memory. A formula for the importance function would be ideal. Sure, such a result is not guaranteed, it all depends on a specific problem.

3. Approximation would be appropriate if it provides significant improvement in computation regardless of whether we manage to approximate the importance function over the whole phase space or not.

For simplicity sake we assume that the albedo of the underlying surface is zero, and there is no absorption. These constraints can easily be removed.

In solving the problem by the Monte Carlo method, we use the "conjugate scheme" (see Section 1.2). Now the problem is reduced to the calculation of the radiation flux leaving the atmosphere of the earth in a given direction (opposite to the solar radiation direction). The radiation source is concentrated at the point S, at which the detector is located. In this case the contribution function φ for the corresponding linear functional I_φ is equal to

$$\varphi(x) = \frac{g(\omega \cdot \omega^*)}{2\pi} \exp(-\tau). \tag{1.15}$$

Here τ is the optical distance between the point r and the atmosphere boundary surface in the direction $(-\omega_\odot)$ of solar radiation.

Assume that the vertical optical thickness of the spherical layer is not large (0.5–1.0). We also omit the observations in which directions of the source are close to the tangent ones. Assume that a particle collides at a point x. Consider in more detail the statistical estimate (4) of the trajectory contribution. If we model the photon trajectories without leaving the layer, then the sum (4) turns into infinite series

$$Q_1 \frac{g(\omega_1 \cdot \omega^*)}{2\pi} \exp(-\tau_1) + Q_2 \frac{g(\omega_2 \cdot \omega^*)}{2\pi} \exp(-\tau_2) + \dots . \tag{1.16}$$

Here τ_i are small on the average. Therefore the statistical fluctuations of the factors $\exp(-\tau_i)$ are also small, compared with the fluctuations of the factors $g(\omega \cdot \omega^*)$. Choose a constant $q < 1$ and replace approximately the values Q_n by q^n (i.e. we assume that the weights decrease as a geometric progression), and replace all the factors $\exp(-\tau_i)$ by 1. Then the series (16) takes the form

$$q \cdot \frac{g(\omega_1 \cdot \omega^*)}{2\pi} + q^2 \cdot \frac{g(\omega_2 \cdot \omega^*)}{2\pi} + \dots . \tag{1.17}$$

Recall that we model the random sequence $\omega_1, \omega_2, \dots$ according to the transition density $g(\omega_i \cdot \omega_{i+1})$. Recall also the form of the Monte Carlo estimate with a given contribution function φ. Now it is easy to see that the series (17) is similar to the statistical estimate for $h(\omega^*)$, where the function $h(\omega)$ is a solution to the integral equation for the functions on the sphere Ω:

$$h(\omega) = q \int\limits_{\Omega} d\omega' \, \frac{g(\omega' \cdot \omega)}{2\pi} \, h(\omega') + \delta_{\omega^*}(\omega). \tag{1.18}$$

But we can obtain the solution to this equation in another way.

First we will show that the function $h(\omega)$, in fact, depends on just one variable $\mu = (\omega \cdot \omega^*)$, i. e.

$$h(\omega) \equiv h(\omega \cdot \omega^*) = h(\mu). \tag{1.19}$$

Let us denote by P the rotation of the sphere Ω around the axis ω^* through an arbitrary angle. Then (19) is equivalent to the statement

$$h(\omega) = h(P\omega).$$

Note that

$$\delta_{\omega^*}(P\omega) \equiv \delta_{\omega^*}(\omega).$$

Substituting $P\omega$ for ω in (18), we obtain

$$h(P\omega) = q \int\limits_{\Omega} d\omega' \, \frac{g(\omega' \cdot P\omega)}{2\pi} \, h(\omega') + \delta_{\omega^*}(P\omega)$$

or

$$h(P\omega) = q \int\limits_{\Omega} d\omega' \frac{g(\omega' \cdot P\omega)}{2\pi} h(\omega') + \delta_{\omega^*}(\omega).$$

Next, for any orthogonal transformation P and for any arbitrary function $f(\omega)$ on the sphere we have

$$P\omega' \cdot P\omega = \omega' \cdot \omega,$$

$$\int\limits_{\Omega} f(P\omega')d\omega' = \int\limits_{\Omega} f(\omega')d\omega'. \tag{1.20}$$

Now put

$$f(\omega') = \frac{g(\omega' \cdot P\omega)}{2\pi} h(\omega')$$

in (20) and use the above results (here the vector ω is fixed and therefore the right-hand side of the equality is a function of ω'). Then we have

$$q \int\limits_{\Omega} d\omega' \frac{g(\omega' \cdot P\omega)}{2\pi} h(\omega') + \delta_{\omega^*}(\omega) = q \int\limits_{\Omega} d\omega' \frac{g(P\omega' \cdot P\omega)}{2\pi} h(P\omega') + \delta_{\omega^*}(\omega)$$
$$= q \int\limits_{\Omega} d\omega' \frac{g(\omega' \cdot \omega)}{2\pi} h(P\omega') + \delta_{\omega^*}(\omega).$$

As a result, we obtain

$$h(P\omega) = q \int\limits_{\Omega} d\omega' \frac{g(\omega' \cdot \omega)}{2\pi} h(P\omega') + \delta_{\omega^*}(\omega).$$

Thus, the function $h(P\omega)$ satisfies the same integral equation as $h(\omega)$ does. And since the equation has the unique solution at $q < 1$, we see that

$$h(P\omega) = h(\omega).$$

Let us expand all the terms of the integral equation (18) into a series in terms of the Legendre polynomials P_k:

$$g(t) = \sum_{k=0}^{\infty} \frac{2k+1}{2} g^k P_k(t),$$

$$h(t) = \sum_{k=0}^{\infty} \frac{2k+1}{2} h^k P_k(t).$$

Here

$$h^k = \int_{-1}^{+1} h(t)P_k(t)dt.$$

To avoid confusion, it is good to bear in mind that unlike g_k the coefficients g^k and h^k are not the nodal values of the functions $g(t)$ and $h(t)$.

By using equation (18), it is easy to obtain simple expressions for the coefficients h^k in terms of g^k. To do this, multiply both sides of (18) by $P_k(\omega \cdot \omega^*)$ in a scalar way and make use of the following formula:

$$\int_\Omega d\omega' P_k(\omega \cdot \omega')P_k(\omega' \cdot \omega'') = \sum_{k=0}^{\infty} \frac{4\pi}{2k+1} P_k(\omega \cdot \omega'')$$

$$\int_\Omega d\omega' f(\omega \cdot \omega')g(\omega' \cdot \omega'') = 4\pi \frac{2k+1}{4}(f, P_k)(g, P_k)P_k(\omega \cdot \omega'').$$

Then we obtain

$$2\pi h^k = 2\pi q h^k g^k + 1,$$

and hence,

$$h^k = \frac{1}{2\pi q h^k g^k + 1}.$$

Now, knowing the coefficients h^k, we can retrieve the function $h(t)$.

Recall that, in fact, we were concerned with the calculation of an approximation to the importance function $f^*(r, \omega)$ and obtained it in the form

$$f^*(r, \omega) \approx h(\omega).$$

Thus, the approximation found does not depend on the space variable r.

Note that we, in fact, have reduced the transfer equation to the simple equation (18) by replacing approximately the exponent of the kernel by a constant $q < 1$. On choosing the constant q see below.

Now consider how the approximate importance function can be used in numerical calculations. First we calculate explicitly the function $h(t)$. To do this, we use the linear approximation with the same nodes μ_i as for the phase function $g(\mu)$. Using the known expansion coefficients for h, we find $h(\mu_i)$, $i = 1, \ldots, n$, and then interpolate the function linearly between the nodes.

Since the obtained expression for the importance function does not depend on the space variable r, we see that in modelling "according to importance" only the probability distribution density of the trajectory directions ω

changes with respect to the "physical" one, while the distribution density of the path lengths remains the same.

The modified probability density of the distribution of the path directions in modelling "according to importance" for the vector ω (at fixed ω' and ω) is of the form

$$\frac{g(\omega \cdot \omega')h(\omega \cdot \omega^*)}{\int\limits_{\Omega} d\omega\, g(\omega \cdot \omega')h(\omega \cdot \omega^*)}. \tag{1.21}$$

Thus, we need to model the probabilistic density of the type (21). There does not exist any suitable algorithm for modelling such density. We can model this density by the universal method of elimination. However, this method has low efficiency if the phase function $g(\mu)$ is highly peaked, i.e. if it increases sharply as $\mu \to 1$.

Another way to model the density is based on approximating the product (21) by the sum

$$p \cdot g(\omega' \cdot \omega) + (1-p) \cdot h(\omega \cdot \omega^*).$$

This approximation is efficient if both of the functions $g(t)$ and $h(t)$ are δ-shaped, i. e. if they increase sharply as $\mu \to 1$. This density is convenient in modelling by randomization, namely we model the vector ω according to the density $g(\omega' \cdot \omega)$ with probability p and according to the density $h(\omega \cdot \omega^*)$ with probability $(1-p)$. The optimal choice of p is considered in Mikhailov (1974). Here the constants norming the density can be easily found in the same manner as for the function $h(t)$.

Now consider how to choose the parameter q. Its value must be approximately equal to the integral estimate of the probability that a particle leaves the atmosphere through either the upper or the lower boundary surface. To estimate q, we assume the following.

1. The altitude h is distributed uniformly over the segment $(0, H)$.

2. The direction ω' of the particle flight is isotropic (i. e. it is distributed uniformly over the sphere Ω).

Recall that here we do not consider the first collisions with fixed ω' since the corresponding contribution can be calculated either analytically or by using a quadrature formula.

Under the above assumptions it is easy to calculate the average probability p that a particle leaves the atmosphere:

$$p = \frac{2}{H} \int_0^{H/2} dh \int_0^{\pi/2} d\theta \frac{1}{2} \left(\exp\left(-\frac{\sigma h}{\cos\theta} \right) + \exp\left(-\frac{\sigma h}{\cos\theta} \right) \right).$$

The sought value is $q = 1 - p$. The integral can be easily calculated numerically.

1.8. MODIFICATION OF RADIATION TRANSFER MODELLING IN SPHERICAL ATMOSPHERE

In the computation of scattered radiation field of a planet the atmosphere is often regarded as a plane slab illuminated from above by parallel rays of solar radiation and bounded below by the Earth's surface. This model is very suitable for computation and gives good accuracy when zenith vision and solar angles do not exceed 75° to 80°. For greater values of these angles this model is not accurate enough. But the use of the spherical model considerably increases computation time.

In this section a modification of trajectory modelling is proposed. This method takes intermediate position between "plane" and "spherical" algorithms.

1.8.1. Statement of the problem

Consider a slab atmosphere model. Atmosphere is divided into horizontally homogeneous slabs. The upper boundary of the atmosphere is illuminated by the solar radiation incident on the upper boundary surface. The problem is to calculate the radiation intensity or another linear functional of the radiation flux.

To model photon trajectories, we should calculate coordinates of the points where a photon intersects the boundaries of layers. For a "plane" model these boundaries are planes, for a "spherical" model those are spheres.

Simple consideration shows that the spherical boundaries may be approximately replaced with plane ones if the thickness of the atmosphere is negligibly small compared with the planet radius, and the angle between the photon direction and the normal vector is not close to 90°.

Coordinates of the intersection point change negligibly in this approximation. The replacement procedure is performed the following way. Let A

be the point of photon collision from which the photon starts. Consider a sphere concentric with the atmosphere boundary and going through the point A. Let the plane P be tangent to the sphere at the point A. Now we replace all the boundary spheres with the planes as shown in Figure 1. Then, modelling a trajectory, we may calculate intersections of the particle with plane boundaries, not spherical ones.

It takes significant effort to change the coordinate system. But we can do the same (and even more than that) in another very simple way described in the next section.

1.8.2. Computational algorithm

Let r and ω be a collision point and flight direction, respectively. Let R be the radius of the boundary sphere that the photon direction $r + l \cdot \omega$ $(l \geq 0)$ intersects first. The distance l from the point r to the intersection point $r' = r + l\omega$ satisfies the equation

$$(r + l \cdot \omega)^2 = R^2.$$

Hence,

$$l = -r\omega \pm \sqrt{(r \cdot \omega)^2 - (r^2 - R^2)}.$$

Here we should choose the least positive root. Consider the case where the flight direction ω intersects the lower layer boundary surface where the particle is located. So, $|r| > R$. Transform the equation for l:

$$l = |r\omega| \left(1 - \frac{r^2 - R^2}{|r\omega|^2}\right)^{1/2}. \tag{1.22}$$

Fix a small ε $(0 \leq \varepsilon \leq 1)$. If

$$\frac{r^2 - R^2}{|r\omega|^2} \leq \varepsilon,$$

then, using the approximate relation $\sqrt{1 - \varepsilon} \approx 1 - \varepsilon/2$, we get

$$\left(1 - \frac{r^2 - R^2}{|r\omega|^2}\right)^{1/2} \approx 1 - \frac{r^2 - R^2}{2|r\omega|^2}.$$

Here the relative error does not exceed $\varepsilon^2/8$. Hence, we obtain an approximate formula for l:

$$l \approx \frac{r^2 - R^2}{|r\omega|},$$

which is more precise than the formula that could be obtained from the consideration in the previous section. Particularly, if we approximately replace $r^2 - R^2$ with $2R(|r| - R)$, then we obtain the precise formula for plane boundary surface:

$$l \approx \frac{|r| - R}{\cos\theta}.$$

Here $r\omega = |r|\cos\theta$, i. e. θ is a zenith angle for the direction ω ($|\omega| = 1$). But in the case

$$\frac{r^2 - R^2}{|r\omega|^2} > \varepsilon$$

we use the precise formula (22).

Now we give a few remarks concerning the proposed development. It is easy to input it into geometrical block of a "standard" program without changing the rest part of the program, which makes the development very suitable in practical computation.

We can obtain a nonmodified "plane" or "spherical" program by setting $\varepsilon = 0$ or $\varepsilon = 1$, respectively.

The computer time savings depend significantly on the number n of layers in the atmosphere model. Numerical experiments have shown that it makes about 20% at $n = 5$, and 50% at $n = 20$ at optical thickness of the atmosphere equal to 0.2–0.3.

Chapter 2.

Generalized transport equation with highly peaked phase function

Radiation transfer in stochastic media can be described by deterministic integral equations of the second kind. When calculating the characteristics of a light field by the Monte Carlo method we obtain a large statistical error if the scattering phase function in this transport equation is highly peaked. In this work we propose statistical estimates and modifications of the modelling process, which allow us to substantially increase the efficiency of calculations. The results obtained may be used to solve three-dimensional problems and allow us to apply other optimizing procedures. We consider the application of the results to the calculation of a flux of scattered solar radiation reflected from a plane layer of broken clouds. We obtain the expressions for modelling densities and give the results of numerical experiments carried out to check the efficiency of the proposed procedure for calculating the flux.

The propagation of radiation in a deterministic medium is described by the transport equation, which is called the standard equation here. The radiation transfer in stochastic media can be described by similar equations, which are obtained from the standard equation by averaging over the ensemble of random realizations of the medium. These equations may be called the generalized transport equations. They describe the probabilistic Markov discrete process of particle transfer in a fictitious homogeneous medium, therefore, they have the form of integral equations of the second kind. These equations differ little from the standard transport equation in integral form. The new parameters may depend in a rather complicated way on the initial physical characteristics of the initial stochastic medium, for example, in the case of a plant canopy or broken clouds (Antyufeev and Marshak, 1988; Glazov and Titiv, 1980). However, the transfer process in this medium, as in the medium considered in the standard transport equation, is described by the density of the free path of a particle between collisions, a scattering phase function (conditional density of a new direction of the free path after scattering), and a source function (the density of the first collisions in the phase space). In practical problems the scattering phase function in the transport equation has highly peaked form. This leads to substantial variations of the statistical error of the unknown quantity, which significantly reduces the accuracy and efficiency of calculations by the Monte Carlo method (we are faced with difficulties in carrying out calculations by other methods as well). As to the standard transport equation in integro-differential form, we can overcome this difficulty if we perform a simple transformation of this equation. Since the flux and collision densities are connected by a simple relation in the standard case, this transformation allows us to optimize the calculations of the linear functionals of the collision density. However, the generalized transport equation may not have the corresponding integro-differential form, and therefore, in this case it is impossible to use the same transformations as in the standard case. In this work we consider the transformations of the integral transport equation in the general form, the modifications of the modelling of the corresponding random process and the estimates of the Monte Carlo method, which allow us to overcome the above difficulty. Throughout the paper we simultaneously consider the standard and general cases to emphasize their similarity and difference. We demonstrate the modifications by the example of one particular calculation. We calculate the angular distribution of solar radiation reflected from a layer of broken clouds, write the model formulas and the estimates of the Monte Carlo method for this problem, give the results of model calculations to demonstrate the efficiency of the methods proposed.

The outline of this chapter is as follows.

In Section 1 we consider the generalized transport equation, the associated probabilistic densities, parameters and notation. We briefly describe the scheme of the Monte Carlo method of calculating the physical characteristics associated with this equation. We consider the difficulties associated with the use of the highly peaked scattering phase function in the Monte Carlo calculations.

In Section 2 we specify a pseudotransport approximation for the highly peaked phase function. Using this approximation, we can construct the optimizing modifications and estimates of the Monte Carlo method.

In Section 3 we give the transformation of the standard integro-differential transport equation. This transformation relies upon the pseudotransport approximation. This transformation is well known and is given here for comparison with the transformations of the generalized (nonstandard) transport equation.

The main results of the work are presented in Section 4, which is divided into three subsections.

In Subsection 4.1 we obtain the estimate of the Monte Carlo method by the pseudotransport approximation. The variance of this estimate is substantially smaller than that of the initial estimate because the highly peaked component of the scattering phase function is eliminated from it.

In Subsection 4.2 we study the transformation of a nonstandard integral transport equation. This transformation relies upon the pseudotransport approximation. To obtain a solution of the transformed equation, we make use of modelling the trajectories in an optically less dense medium with a smaller number of collision events of a particle before its escape from the region.

In Subsection 4.3 we consider the problem of calculating the flux for the generalized transport equation. We have obtained the result analogous to that obtained in Section 3.

In Section 5 we apply the results obtained to the particular problem of calculating a flux of scattered solar radiation, which is reflected from a plane layer of broken stochastic clouds.

In Subsection 5.1 we obtain the formulas for the parameters, which specify the modified density when modelling the path length and for the contribution function, which may be used when estimating the error in the Monte Carlo calculations for the flux of reflected radiation.

In Subsection 5.2 we consider the results of model calculations. In conclusion we make some remarks about the application of the results to the solution of direct and inverse problems.

2.1. GENERALIZED TRANSPORT EQUATION

The phase density of collisions $F(x)$ satisfies the generalized integral equation

$$F(x) = \int_X k(x', x)F(x')dx' + h(x) \tag{2.1}$$

with the kernel

$$k(x',x) = \frac{\lambda g(\omega',\omega) f(|r-r'|)}{|r-r'|^2}\delta_s(\omega),$$

where

$$s = \frac{r-r'}{|r-r'|}.$$

In the following we use the notation:

$X = R \times \Omega$ is the phase space of the equation, which consists of points $x = (r, \omega)$;

R is the geometric space of collision points $r = (x, y, z)$ with the Cartesian coordinates;

Ω is the space of unit vectors $\omega = (a, b, c)$ of the direction of particles immediately before their collision;

$g(\omega', \omega)$ is the scattering phase function. In fact, in the standard case g is a function of a single variable and depends only on the scalar product $\omega' \cdot \omega$. In the generalized equation the scattering phase function is a function of two unit vectors ω' and ω. For simplicity we do not consider here the function g versus the space coordinate r. In fact, the results can be extended without difficulty to this case as well;

f is the probabilistic density of the free path of a photon between two successive collisions. This function is specified on rays of the form

$$\{r : r(l) = r' + l \cdot \omega, \ \ l \geq 0\},$$

where r' is the initial point of the particle path. In the standard case the function f has the special form:

$$f(l) = \sigma(r(l)) \exp(-\tau(l)),$$

where

$$\tau(l) = \int_0^l \sigma(r(t))dt$$

is the optical length of the free path. This density may have arbitrary form in the generalized equation;

$h(x)$ is the density of points x at the first collisions in the phase space. The first collisions are usually modelled in the following way. We consider the space probabilistic density $h_R(r)$ of a physical radiation source and model the point r according to this density. Then we choose the direction ω according to the density $h_\Omega(\omega)$. Finally we model the "zero" path length according to the density $f(l)$ and calculate the point r_1 of the first collision. In the generalized equation the "zero" path length may be chosen according to some other density rather than the density $f(l)$. For example, this is the case of a cloud layer (see Glazov and Titiv (1980) and Section 6 in this chapter);

λ is the probability of the photon survival at a collision. For simplicity we consider only the case in which there is no absorption ($\lambda = 1$).

The absorption can readily be taken into account and the corresponding formulas are given without proof.

Many of the physical characteristics of the scattered radiation field may be represented as linear functionals

$$(F, \varphi) = \int_X F(x)\varphi(x) \tag{2.3}$$

of the collision density F, which is a solution of (1), and the function φ is dependent on the sought physical characteristic. In order to calculate (3) we use the collision estimate of the Monte Carlo method (Spanier and Gelbard, 1969). We model the Markov chain $x_1, x_2, \ldots, x_n$ of successive collisions of a particle in the phase space X. Here n is the number of the last collision of the particle before it escapes from the region (or it is absorbed if absorption is taken into account). This process of random walk of a particle is completely specified by the functions f, g, h. As a result of the modelling we obtain a sample according to the collision density $F(x)$. At each collision point x we calculate the value of the contribution function $\varphi(x)$. The sum

$$\varphi(x_1) + \varphi(x_2) + \ldots + \varphi(x_n) \tag{2.4}$$

yields a statistical estimate of the contribution of the entire trajectory $x_1, x_2, \ldots, x_n$. Finally, all these contributions are averaged over the number of trajectories.

In practical calculations we often use the highly peaked scattering phase function (i.e. the phase function that abruptly increases as $\mu \to 1$; this concept is more precisely defined in Section 2). If the scattering phase function $g(\omega', \omega)$ enters as a cofactor into the expression for the contribution function φ, significant variations of its values lead to significant variations of the estimate of the contributions from the collisions and the trajectories. For example, in the case of a cloud medium the scattering phase function $g(\omega', \omega)$ is highly peaked. The values of the phase function at one point may be thousands of times greater than those at the other point. Besides, a cloud layer generally has large optical thickness. Therefore, a particle suffers, on the average, tens and hundreds of collisions before it escapes from the layer. The modelling of long trajectories, naturally, requires a lot of computation time. In Section 5 we consider the modifications of the standard calculation scheme, when the scattering phase function is highly peaked. Due to these modifications we can overcome the above difficulties and substantially increase the efficiency of calculations by the Monte Carlo method. The idea of all the methods is to replace the initial phase function by an approximation, which is called a "pseudotransport" approximation.

2.2. APPROXIMATION FOR HIGHLY PEAKED PHASE FUNCTION

For simplicity we first consider the one-dimensional case in which the phase function depends on the scalar product $\mu = \omega' \cdot \omega$. The term "highly peaked" implies that the function $g(\mu)$ has a specific form — it dramatically increases as the argument $\mu \to 1$. In this case we can choose a small positive number ε and divide the interval $(-1, +1)$, where the phase function $g(\mu)$ is defined, into two subintervals: $(-1, 1-\varepsilon)$ and $(1-\varepsilon, 1)$. The large values of the function $g(\mu)$ are collected on the small subinterval $(1-\varepsilon, 1)$, and the values of this function do not vary significantly on the subinterval $(-1, 1-\varepsilon)$ (see Figure 4). This function can be represented as a linear combination of the highly peaked function g_ε and the "regular" function $\widetilde{g}$

$$g = pg_\varepsilon + q\widetilde{g}, \tag{2.5}$$

$$p + q = 1, \quad p > 0, \ q > 0,$$

where the coefficients p and q are equal to

$$p = \int_\varepsilon^1 g(\mu)d\mu,$$

$$q = \int_{-1}^\varepsilon g(\mu)d\mu,$$

and p is comparable to q in oder of magnitude. The functions g_ε and $\widetilde{g}$ are specified as follows:

$$g_\varepsilon(\mu) = \begin{cases} \dfrac{1}{p}\, g(\mu), & \mu > 1-\varepsilon, \\ 0, & \mu \le 1-\varepsilon, \end{cases}$$

$$\widetilde{g}(\mu) = \begin{cases} \dfrac{1}{q}\, g(\mu), & \mu \le 1-\varepsilon, \\ 0, & \mu > 1-\varepsilon. \end{cases}$$

We choose the value ε ($0 < \varepsilon \ll 1$) arbitrarily. Note that this representation (5) is chosen so that the functions $\widetilde{g}$ and g_ε are probabilistic densities, and $p + q = l$. Consequently, we can model the density g as a "mixture" of the densities $\widetilde{g}$ and g_ε. We consider the function $g_\varepsilon(\mu)$ as a regular approximation to the distribution $\delta(\mu - 1)$. Then

$$\int\limits_\Omega g(\omega'\cdot\omega) f(\omega')d\omega' = p\int\limits_\Omega g_\varepsilon(\omega'\cdot\omega) f(\omega')d\omega' + q\int\limits_\Omega f(\omega')d\omega'$$
$$\approx pf(\omega) + q\int\limits_\Omega f(\omega')d\omega'. \tag{2.6}$$

This implies that the function g may approximately be replaced by the distribution

$$g_{tr} = p\delta_1 + q\widetilde{g}. \tag{2.7}$$

The function g_{tr} is called the "pseudotransport approximation" of the phase function g. This approximation has the form of the phase function $g(\mu)$ on the interval $(-1, 1-\varepsilon)$. It is more precise than the transport approximation, where instead of the highly peaked phase function of the general form we consider a probabilistic mixture of the delta function concentrated at the point 1 and an isotropic scattering phase function. In the general case the scattering phase function $g(\omega', \omega)$ is a function of two vectors ω' and ω. It is said to be highly peaked if (for fixed ω) it dramatically increases as $\omega' \to \omega$. We can choose ε: $0 < \varepsilon \ll 1$ such that the integrals

$$p(\omega) = \int\limits_{\Omega_\varepsilon} g(\omega'\cdot\omega)d\omega',$$

$$q(\omega) = \int\limits_{\overline{\Omega}_\varepsilon} g(\omega'\cdot\omega)d\omega'$$

are comparable in order of magnitude (even though ε is small). Here

$$\Omega_\varepsilon = \{\omega : \omega' \cdot \omega \leq \varepsilon\},$$

$$\overline{\Omega}_\varepsilon = \{\omega : \omega' \cdot \omega > \varepsilon\}.$$

In practice this implies that if a particle travels in the direction ω', it is scattered with a sufficiently great probability in the direction ω, which is close to the initial direction ω'.

As in the one-dimensional case (7), the "pseudotransport approximation" in the general case is

$$g_{tr}(\omega', \omega) = p(\omega')\delta(\omega', \omega) + q(\omega')\delta(\omega', \omega). \tag{2.8}$$

Here

$$\widetilde{g}(\omega' \cdot \omega) = \begin{cases} \dfrac{1}{q}\, g(\omega' \cdot \omega), & \omega \in \overline{\Omega}_\varepsilon, \\ 0, & \omega \notin \overline{\Omega}_\varepsilon. \end{cases}$$

and the function $\widetilde{g}(\omega' \cdot \omega) \equiv \widetilde{g}(\omega', \omega)$ of the variable ω (for fixed ω') is a probabilistic density on the sphere Ω.

We have obtained the optimizing modifications of the standard calculation procedure, which allows us to decrease the variations of the contribution function and increase the efficiency of calculations. All the modifications are based on the substitution $g \to g_{tr}$. We replace the phase function by its "transport approximation". As we see below, this substitution allows us to reduce the problem of the highly peaked (irregular) scattering phase function g to the problem of the phase function $\widetilde{g}$, in which the variations of its values are much smaller. We first consider the results of this substitution for the standard transport equation and then apply this transformation to the case of the general equation.

2.3. TRANSFORMATION OF STANDARD TRANSPORT EQUATION

The substitution $g \to g_{tr}$ in the standard transport equation immediately yields better results. We should only consider the transport equation for a radiation flux in the integro-differential form:

$$(\omega \cdot \operatorname{grad} \Phi(r, \omega)) + \sigma(r)\Phi(r, \omega)) = \sigma(r) \int\limits_{\Omega} \Phi(r, \omega) g(\omega' \cdot \omega) d\omega' + \Phi_0(r, \omega).$$

Here $\sigma(r)$ is the scattering cross-section. Recall that we consider a nonabsorbing medium, therefore, there is no absorption coefficient in the equation.

On the substitution $g \to g_{tr}$ the equation has the form

$$\omega \cdot \nabla\Phi(r,\omega) + \sigma(r)\Phi(r,\omega) \\ = p\sigma(r)\Phi(r,\omega) + q\sigma(r)\int\limits_{\Omega}\Phi(r,\omega)\widetilde{g}(\omega'\cdot\omega)d\omega' + \Phi_0(r,\omega) \tag{2.9}$$

or, after transformation,

$$\omega \cdot \nabla\Phi(r,\omega) + q\sigma(r)\Phi(r,\omega) = q\sigma(r)\int\limits_{\Omega}\Phi(r,\omega)\widetilde{g}(\omega'\cdot\omega)d\omega' + \Phi_0(r,\omega).$$

The idea of this transformation is most probably suggested in Candelore and Gast (1964). We have obtained the transport equation for the medium with the new scattering cross-section $\widetilde{\sigma} \equiv q\sigma$ and the new scattering phase function $\widetilde{g}$. Since the function $\Phi(r,\omega)$ remains unchanged in the equation, we can state that the value of the flux does not change (to be more exact, it changes insignificantly) if we model the particle trajectories in the medium with the parameters $\widetilde{\sigma}$ and $\widetilde{g}$ instead of the initial parameters σ and g. The advantage of this technique is that the new medium is optically less dense than the initial medium, and the number of collisions in the lifetime of a single particle decreases. Besides, the variation of the values of the new phase function $\widetilde{g}$ is much smaller than that of the initial phase function g.

This result resembles an algebraic trick, but it will be explained in Section 6. The transformation is simple because the probabilistic density of the free path has a special form, viz. it is an exponential density.

As mentioned above, the form of the generalized integral equation (1) differs insignificantly from the standard transport equation in integral form. However, there is an essential difference: the standard transport equation can be written in integro-differential form, whereas equation (1) cannot be written in this form. Therefore, we have to consider the transformations of the integral equation (1). The necessary transformations become more complicated and the final formulas more cumbersome.

2.4. OPTIMIZING PROCEDURES FOR MODELLING AND CALCULATION

In this section we consider three different modifications of the standard computational method, using the substitution $g \to g_{tr}$. As mentioned above, we consider the generalized transport equation in integral form. We may try to transform this equation by the direct substitution $g \to g_{tr}$, as it was done in the standard case for the integro-differential equation. However, we have not made progress in this direction. We see below that the algebraic transformations alone cannot yield the necessary result. In order to understand how to transform the integral equation, we should first study in detail the modelling of the particle trajectories when their direction changes according to the "pseudotransport" phase function. We will deal with this case in Subsection 4.3 but we first consider the transformation of a local collision estimate.

2.4.1. Modification of the local estimate

In problems of estimating linear functionals of the form (F, φ) the function $g(\mu)$ enters as a cofactor into the expression for the contribution function φ in the Monte Carlo collision estimate. In particular, if we need to calculate the collision density at a point of the phase space (r, ω^*) or the flux of scattered radiation in the direction ω^*, we can represent φ as

$$\varphi(r, \omega) = g(\omega' \cdot \omega) \cdot \varphi_1(r, \omega), \tag{2.10}$$

where the function φ no longer depends on the values of the phase function g. In this case the statistical error of the calculation is closely connected with the variations of the values of the phase function $g(\mu)$. The substitution $g \to g_{tr}$ allows us to get rid of the highly peaked component g_ε of the phase function g and thus improve the corresponding estimate of the Monte Carlo method. Now we consider the expression for the linear functional

$$(F, \varphi) \equiv \int_\Omega d\omega \int_R dr F(r, \omega)\varphi(r, \omega)$$

of the solution F of the transport equation (1) and transform it into the form, which is convenient for estimating it by the Monte Carlo method. We first replace the function $\varphi(r, \omega)$ by its expression (10):

$$\int_\Omega d\omega \int_R dr F(r, \omega)\varphi(r, \omega) = \int_\Omega d\omega \int_R dr F(r, \omega) g(\omega, \omega^*) \cdot \varphi_1(r, \omega)$$

$$= \int_\Omega d\omega \; g(\omega, \omega^*) \left(\int_R dr F(r, \omega)\varphi_1(r, \omega) \right).$$

Further we use the pseudotransport approximation $g \to g_{tr}$:

$$\int_{\Omega} d\omega\ g(\omega,\omega^*)\left(\int_{R} drF(r,\omega)\varphi_1(r,\omega)\right)$$
$$\simeq \int_{\Omega} d\omega\ g_{tr}(\omega,\omega^*)\left(\int_{R} drF(r,\omega)\varphi_1(r,\omega)\right)$$
$$= p\int_{R} drF(r,\omega^*)\varphi_1(r,\omega^*) + q\int_{\Omega} d\omega \int_{R} drF(r,\omega)(\tilde{g}(\omega'\cdot\omega^*)\varphi_1(r,\omega)).$$

The second term has the form, which is convenient for estimating it by the Monte Carlo method. Here F is the modelled density of collisions, and $\tilde{g}(\omega'\cdot\omega^*)\varphi_1(r,\omega)$ is a regular contribution function (the values of the highly peaked component of the phase function are eliminated from it). We transform the integral in the first term. We can consider equation (1) as an integral expression for $F(r,\omega^*)$:

$$F(r,\omega^*) = \int_{\Omega} d\omega' \int_{R} dr' k((r',\omega'),(r,\omega^*))F(r',\omega') + h(r,\omega^*).$$

Substituting this expression for $F(r,\omega^*)$ under the integral, we obtain

$$\int_{R} dr\ F(r,\omega^*)\varphi_1(r,\omega^*)$$
$$= \int_{R} dr\left(\int_{\Omega} d\omega' \int_{R} dr' F(r',\omega')k((r',\omega'),(r,\omega^*)) + h(r,\omega^*)\right)\varphi_1(r,\omega^*)$$
$$= \int_{\Omega} d\omega' \int_{R} dr' F(r',\omega')\left(\int_{R} dr\ k((r',\omega'),(r,\omega^*))\varphi_1(r,\omega^*)\right)$$
$$+ \int_{R} dr\ h(r,\omega^*)\varphi_1(r,\omega^*).$$

In the first term r', ω' and r are related variables, since the integral is taken with respect to them. Therefore, we can change the notation and rewrite the integral as

$$\int_{\Omega} d\omega' \int_{R} dr' F(r',\omega') \left(\int_{R} dr\ k((r',\omega'),(r,\omega^*))\varphi_1(r,\omega^*) \right)$$
$$= \int_{\Omega} d\omega \int_{R} dr\ F(r,\omega') \left(\int_{R} dr'\ k((r,\omega),(r',\omega^*))\varphi_1(r',\omega^*) \right)$$
$$= \int_{R} dr' \left(\int_{R} dr \int_{\Omega} d\omega\ F(r,\omega')\, k((r,\omega),(r',\omega^*)) \right) \varphi_1(r',\omega^*).$$

The last integral is represented in the form, which is convenient for calculating it by the Monte Carlo method. Indeed, we can consider the expression between brackets as the modelling density and $\varphi_1(r',\omega^*)$ as a regular contribution function. Let us consider the expression between brackets in greater detail. It is a superposition of two probabilistic densities $F(r,\omega)$ and $k((r,\omega'),(r',\omega^*))$. Using this density, we model successively the point (r,ω) and then a transition from (r,ω) to (r',ω^*). It remains to calculate the regular contribution function $\varphi_1(r',\omega^*)$.

The above modification of the local estimate is suitable because it does not change the modelling algorithm and may be used to simultaneously calculate several functionals. Finally, it is no obstacle to the application of other optimizing procedures.

2.4.2. Modification of trajectory modelling

We rewrite the transport equation (1) in an extended form:

$$F(r,\omega) = \int_{\Omega} d\omega'\ g(\omega',\omega) \int_{R} dr' \frac{g(\omega',\omega) f(|r-r'|)}{|r-r'|^2} \delta_s(\omega) F(r',\omega') + h(r,\omega). \tag{2.11}$$

We consider the first term in the right-hand side of this equation. We transform the integral taken with respect to the variable r. We designate

$$l = |r'-r|, \qquad s = \frac{r-r'}{|r-r'|}, \qquad |s| = 1, \qquad r' = r - l \cdot s$$

and change the variables in this integral: we pass from the Cartesian (x,y,z) to the polar coordinates (l,s) $(0 \le l < \infty,\ s \in \Omega)$. The Jacobian of the transformation

$$T:\ (x,y,z) \to (l,s)$$

is equal to l^2. This transformation allows us to get rid of the delta function in the kernel of the integral operator in (10). We rewrite the integral as

$$\int_R dr' \frac{g(\omega',\omega)f(|r-r'|)}{|r-r'|^2}\delta_s(\omega)F(r',\omega')$$

$$= \int_\Omega ds \int_0^\infty dl\; l^2 \frac{f(l,\omega)}{l^2}\delta_s(\omega)F(r-l\cdot s,\omega')$$

$$= \int_\Omega ds\; \delta_s(\omega) \int_0^\infty dl\; f(l,\omega)F(r-l\cdot s,\omega')$$

$$= \int_0^\infty dl\; f(l,\omega)F(r-l\cdot\omega,\omega'). \tag{2.12}$$

It is a one-dimensional integral over the ray

$$L = \{r' : r' = r - s\cdot\omega\}, \quad s \geq 0,$$

with its origin at the point r and the direction $-\omega$. The variable l corresponds to the variable r' here. The point r in (10) is also variable, however, the fixed value of the parameter l ($l = 0$) corresponds to it in the chosen reference system. We introduce the parameter t independent of the point r. We extend the ray L beyond the initial point r, choose, and fix an arbitrary point r_0 on this extension ray, e.g. a point at the boundary of the domain (see Figure 2). By the definition of r_0 the points r and r' lie on the ray

$$K = \{r_0 + t\cdot\omega : t < 0\}.$$

Let

$$r = r_0 + t\cdot\omega, \qquad r' = r_0 + t'\cdot\omega \qquad (t,t' < 0).$$

In the following transformations we sometimes have to consider the functions F and f for a fixed ω unless otherwise stated. Therefore, in the notation of these functions we use ω as an index.

Besides, instead of the vector arguments $r_0 - t\cdot\omega$ and $r_0 - t'\cdot\omega$ we use the corresponding scalar quantities t and t' and denote the functions of the new variables by the same letter. Using this notation, we rewrite the integral (11) as

$$\int_0^\infty dl\; f(l,\omega)F(r-l\cdot\omega,\omega') = \int_{-\infty}^t dt'\; f_\omega(t-t')F_{\omega'}(t').$$

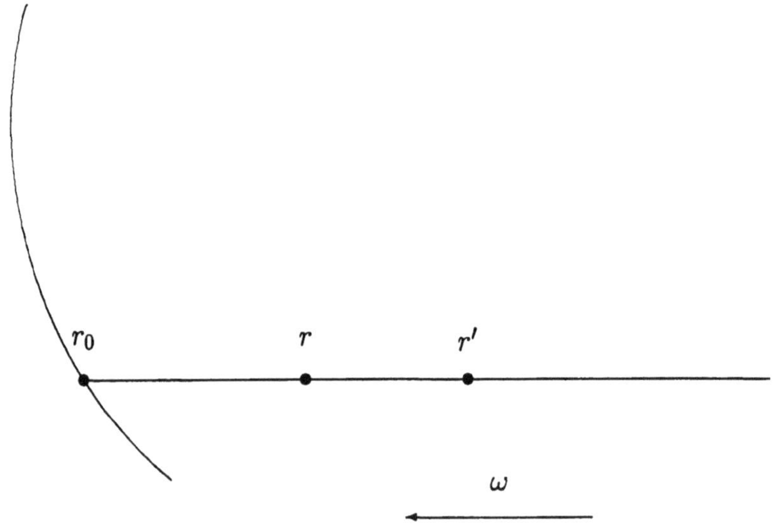

Figure 2: Scheme of the location of the points r_0, r, and r'

Taking into account that $f_{\omega'}(l) = 0$ for $l < 0$, we can rewrite this expression as

$$\int_{-\infty}^{t} dt'\ f_\omega(t-t')F_{\omega'}(t') = \int_{-\infty}^{+\infty} dt'\ f_\omega(t-t')F_{\omega'}(t') \equiv [f_\omega * F_{\omega'}](t).$$

Here $[f_\omega * F_{\omega'}](t)$ is the convolution of the functions f_ω and $F_{\omega'}$ of a single variable. Finally, we can rewrite (10) as

$$F_\omega(t) = \int_{\Omega} d\omega'\ g(\omega', \omega) \cdot [f_\omega * F_{\omega'}](t) + h_\omega(t). \qquad (2.13)$$

If we apply the Laplace transform with respect to the variable t to both sides of equation (10), the convolution of the functions becomes a product and the equation has the form

$$\widehat{F}_\omega(s) = \int_{\Omega} d\omega'\ g(\omega', \omega) \cdot \widehat{f}_\omega(s)\widehat{F}_{\omega'}(s) + \widehat{h}_\omega(s).$$

Here the sign ^ over the symbol of a function implies its Laplace transform. The substitution $g \to g_{tr}$ yields

$$\widehat{F}_\omega(s) = \int\limits_\Omega d\omega'\; g_{tr}(\omega', \omega) \cdot \widehat{f}_\omega(s)\widehat{F}_{\omega'}(s) + \widehat{h}_\omega(s),$$

or

$$\widehat{F}_\omega(s) = p_\omega \widehat{f}_\omega(s)\widehat{F}_\omega(s) + q_\omega \int\limits_\Omega d\omega'\; \widetilde{g}(\omega', \omega) \cdot \widehat{f}_\omega(s)\widehat{F}_{\omega'}(s) + \widehat{h}_\omega(s).$$

Only now can we perform simple algebraic transformations similar to those performed with the standard integro-differential equation in Section 4.

If we transpose the term $p_\omega \widehat{f}_\omega(s)\widehat{F}_\omega(s)$ to the left-hand side and divide both sides of the equation by $(1 - p_\omega \widehat{f}_\omega(s))$, we obtain successively

$$\widehat{F}_\omega(s)(1 - p_\omega \widehat{f}_\omega(s)) = \int\limits_\Omega d\omega'\; \widetilde{g}(\omega', \omega) \cdot q_\omega \widehat{f}_\omega(s)\widehat{F}_{\omega'}(s) + \widehat{h}_\omega(s),$$

$$\widehat{F}_\omega(s) = \int\limits_\Omega d\omega'\; \widetilde{g}(\omega', \omega) \cdot \frac{q_\omega \widehat{f}_\omega(s)}{1 - p_\omega \widehat{f}_\omega(s)} \widehat{F}_{\omega'}(s) + \frac{1}{q_\omega} \cdot \frac{q_\omega \widehat{h}_\omega(s)}{1 - p_\omega \widehat{f}_\omega(s)}.$$

Finally, if we apply the inverse Laplace transform to both sides of the last equation, we obtain

$$F_\omega(t) = \int\limits_\Omega d\omega'\; \widetilde{g}(\omega', \omega) \cdot [\widetilde{f}_\omega * F_{\omega'}](t) + \frac{1}{q_\omega} \cdot \widetilde{h}_\omega(t). \tag{2.14}$$

Here $\widetilde{f}$ and $\widetilde{h}$ are the preimages of the Laplace functions

$$\frac{q_\omega \widehat{f}_\omega}{1 - p_\omega \widehat{f}_\omega} \quad \text{and} \quad \frac{q_\omega \widehat{h}_\omega}{1 - p_\omega \widehat{f}_\omega},$$

respectively. We can immediately prove that the functions $\widetilde{f}$ and $\widetilde{h}$ are probabilistic densities, but we do not consider this here. We deal with these functions in the next section, where this fact becomes obvious. We multiply both sides of (13) by q_ω:

$$q_\omega F_\omega(t) = \int\limits_\Omega d\omega'\; \widetilde{g}(\omega', \omega) \cdot [\widetilde{f}_\omega * q_\omega F_{\omega'}](t) + \widetilde{h}_{\omega(t)}. \tag{2.15}$$

The form of this equation is completely identical to that of the transformed equation (12). Equation (14), if necessary, can be reduced by inverse transformations to the same form as equation (10), only the functions f, g and h are replaced by $\widetilde{f}$, $\widetilde{g}$ and $\widetilde{h}$ and the collision density F by qF. This implies that if the phase function $\widetilde{g}$, the initial collision density $\widetilde{h}$, and the density $\widetilde{f}$ of the free path length are used when modelling the trajectories, we obtain the collision density qF instead of F, where $q < 1$, i. e. we have obtained the new medium, which is optically less dense than the initial one. Instead of N collisions we model, on the average, qN collisions, i. e. we save on the modelling of collisions. Naturally, the final result of the calculations should be divided by q.

2.4.3. Modifications of the calculation of the radiation flux

In this section we show how we can modify the procedure for calculating the flux by the Monte Carlo method if the phase function is highly peaked. The procedure proposed is very similar to that discussed in Section 3 and incorporates all the advantages obtained in Subsections 4.1 and 4.2. In order to perform this modification we need to consider the procedure for modelling the trajectory in greater detail.

We first model the point x_1 of the first collision. By equation (1), we should model this point according to the density $h(x_1)$. However, in practice we model the space point r_0 of a physical source (it can sometimes be determined), then the initial direction ω_1, and, finally, the point x_1. We model a random free path according to the density $h_l(t)$ (generally speaking, $h_l \neq f$) and then find the point $r_1 = r_0 + \omega_1 \cdot l$ (the direction ω_1 has already been specified).

Further we model successively the transition $\omega_1 \to \omega_2$ of the direction of flight according to the probabilistic density $g(\omega', \omega)$, the free path according to the density $f(l, \omega)$, and so on. If a successive modelled collision point is inside the layer, we include the contribution in the statistical estimate of the functional of the trajectory. Otherwise we stop modelling this trajectory and start modelling the succeeding one.

We now consider the peculiarities of modelling a photon trajectory, when we use the phase function g_{tr}.

Let r' be a space point of a successive collision and ω' be the direction of a particle flight immediately before it hits r'. We should now model the direction of a particle flight after it is scattered at r' according to the density $g_{tr}(\omega', \omega)$ (for fixed ω'). This density (see Section 2) is a probabilistic mixture of the delta distribution $\delta_{\omega'}(\omega)$ and the regular density $\widetilde{g}(\omega', \omega)$. Therefore,

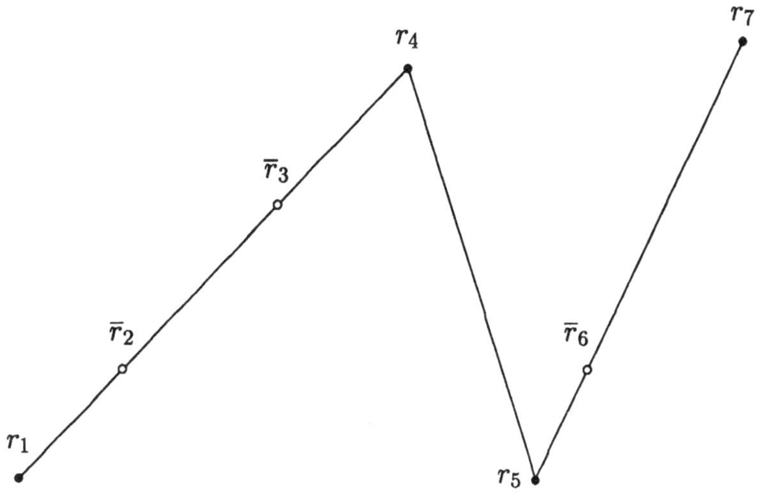

Figure 3: Interval of the particle trajectory with "real" and delta collisions

the algorithm for modelling the direction ω is as follows. We put $\omega = \omega'$ with probability $p_{\omega'}$ (i. e. the direction does not change), and the vector ω should be modelled with probability $q_{\omega'}$ according to the density $\widetilde{g}(\omega', \omega) \equiv \widetilde{g}(\omega', \omega)$.

We now consider a fixed trajectory of a particle. We divide the space collision points r_i, $i = 1, 2, \ldots, n$, into two groups. The point r_0 and all points r_i belong to group 1, at these points the particle changes its direction of flight, i. e. the direction ω is modelled according to the phase function $\tilde{g}$. These points are called the "true" scattering points. All the other space scattering points belong to group 2, i. e. the points at which a particle does not change its direction after being scattered. This scattering is commonly called delta scattering. These points are called intermediate scattering points and denoted by $\overline{r}_i$. Figure 3 shows such a trajectory. When modelling the trajectory we may try to neglect the intermediate scattering points and model only the "true" scattering points. To do this, we must learn to model the free path between the "true" scattering points.

Let us fix a "true" scattering point r different from r_0 (we consider r_0 later). Let ω be the direction of a particle flight immediately after scattering at the point r. Below we only deal with the direction ω, therefore, for simplicity we use designations p, q instead of p_ω, q_ω. We denote by ξ_1 a random path length from the point r to the succeeding collision point (it makes no difference whether it is a "true" or intermediate one). The particle experiences the "true" scattering, changes its flight direction with probability q, and continues moving in the initial direction ω with probability p. In the latter case, on travelling a distance ξ_2, the particle experiences collision, changes its direction with probability q, and continues moving in the initial direction with probability p, and so on.

Our main interest here is with the probabilistic distribution of the path length l in the initial direction ω, viz. the density of this probabilistic distribution denoted by $\tilde{f}$.

All random quantities $\xi_1, \xi_2, \dots$ have the same probabilistic distribution with density $f(t)$ and are mutually independent. The random path length l in the initial direction ω is distributed as follows. It takes the value

$$l_1 = \xi_1$$

with probability q (i. e. the first scattering turns out to be a "true" one with probability q), the value

$$l_1 = \xi_1 + \xi_2$$

with probability qp (i. e. the first scattering turns out to be an intermediate one with probability p, and the second scattering turns out to be a "true" one with probability q), the value

$$l_3 = \xi_1 + \xi_2 + \xi_3$$

with probability qp^2 (i. e. the first two scatterings turn out to be intermediate ones with probability p^2, and the third scattering turns out to be a "true" one with probability q), and so on. We denote the densities of the probabilistic distributions of random quantities $l_1, l_2, \dots$ by $f_1, f_2, \dots$ (here $f_1 \equiv f$). Since the density of the sum of independent random quantities is equal to the convolution of their densities, we have

$$\begin{aligned} f_1 &\equiv f, \\ f_2 &= f * f, \\ f_3 &= f * f * f \end{aligned}$$

and so on. Hence, the sought probabilistic density $\tilde{f}$ for the path length l is

$$\tilde{f} = q \cdot f_1 + qp \cdot f_2 + qp^2 \cdot f_3 + \dots .$$

If we apply the Laplace transform to this relation and denote the Laplace transforms of the functions $\tilde{f}$, f_1, f_2, ... by $\widehat{(\tilde{f})}$, $\hat{f}_1$, $\hat{f}_2$, ..., then we obtain the geometric series

$$\widehat{(\widetilde{f})} = q \cdot \widehat{f} + qp \cdot \widehat{f}^2 + qp^2 \cdot \widehat{f}^3 + \ \ldots,$$

whose sum is readily calculated:

$$\widehat{(\widetilde{f})} = \sum_{k=0}^{\infty} qp^k \widehat{f}^{k+1} = \frac{q\widehat{f}}{1 - p\widehat{f}}.$$

If we apply the inverse Laplace transform to both sides of the last relation, we obtain the formula for the sought density $\widetilde{f}$, viz.

$$\widetilde{f} = \mathcal{L}^{-1}\left(\frac{q\widehat{f}}{1 - p\widehat{f}}\right), \tag{2.16}$$

where $\mathcal{L}^{-1}$ is the inverse Laplace transform. Recall that we have obtained the probabilistic density of the distribution of a random path between two successive points of "true" scattering, the initial point being different from r_0. We proceed to the case in which the initial point of the path is r_0. In this case we repeat all the above reasoning. There is only one difference, viz. the probabilistic distribution of the random path length ξ_1 is specified by the function $h_l(t)$ rather than $f(t)$. Accordingly, the expressions for the density functions f_1, f_2, ... have the form:

$$f_1 \equiv h_l,$$
$$f_2 = h_l * f,$$
$$f_3 = h_l * f * f,$$

and so on. The final formula is

$$\widetilde{h}_l = \mathcal{L}^{-1}\left(\frac{q\widehat{h}_l}{1 - p\widehat{f}}\right). \tag{2.17}$$

Note that we have obtained the same functions (15) and (16) in Subsection 5.2. Thus, we see that they are probabilistic densities. Besides, it has become clear why we apply the Laplace transform to transform the integral transport equation. In Section 6, where we deal with analytic formulas, we see why the corresponding transform of the integro-differential transport equation is readily performed. Thus, if we want to model particle trajectories neglecting the intermediate points, at which the particles experience delta scattering, we should use the probabilistic densities $\widetilde{f}$ and $\widetilde{h}$ to model the free path length between the collisions and the phase function $\widetilde{g}$ to model the direction of scattering. As shown in Subsection 4.2, we can reduce the number of modelled collisions if we use the densities $\widetilde{f}$ and $\widetilde{h}$ of the path; in this case the collision density function is multiplied by the constant $q < 1$. Moreover, we can make the same conclusion from the above reasonings. Indeed, let N be the average number of collision points in the trajectory (both "true" and intermediate ones). Each collision point is a delta scattering point with probability p and it is a "true" scattering point with probability q. Thus, and if we neglect the delta scattering points, the average number of points in the trajectory decreases and equals qN. We have obtained the same result in Subsection 4.2.

However, although the collision density has changed, the modelled flux remains unchanged. Indeed, it is obvious that in the new problem, as compared to the original one, the rectilinear sections of the trajectory remain unchanged. Only the intermediate collision points disappear on the line segment between the two points at which a particle changes its flight direction (Figure 3). Consequently, the number of particles, which cross an arbitrary area in a given direction remains unchanged. Thus, the flux density does not change, i. e. the flux density for the new model of the medium specified by the functions $\widetilde{h}$, $\widetilde{f}$ and $\widetilde{g}$ is the same as that for the initial model specified by the functions h, f and g. Therefore, we can calculate the initial flux density, using the Monte Carlo method for the new model of the medium, where radiation transfer takes place. In this case we model the direction of scattering according to the phase function $\widetilde{g}$ and the path length according to the densities $\widetilde{h}$ and $\widetilde{f}$, but we should replace the initial contribution function $\varphi(r, \omega)$ by $\widetilde{\varphi}(r, \omega)$ and the initial medium parameters by the new ones. We consider the corresponding example for the particular problem in Subsection 5.1.

Let us see how we benefit from this modification. The number of modelled collisions before the particle escape reduces, which results in reducing the time of modelling the trajectories. The spread in the values of the phase function $\widetilde{g}$ is much smaller than that of the initial function g. Therefore, the statistical spread in values of the trajectory contribution also decreases, which increases the accuracy of calculations. This result agrees very closely with that obtained in Section 3 for the integro-differential transport equation.

Recall that so far we have considered the scattering medium without considering absorption. It remains to write the formulas in which absorption is taken into account.

If in the above reasonings we also consider the possibility that the particles are absorbed at collisions, of course the derivation of the formulae will become more difficult than that for the nonabsorbing case. However, we may act other way. We introduce the "weight" of a particle. In fact, this "weight" was implicitly present in the above reasonings but was equal to unity. When there is absorption, i. e. $\lambda < 1$, only the initial "weight" of a particle at the initial instant is unity. Further the weight W changes discretely from collision to collision. A successive value of the weight W_{i+1} is determined using the preceding weight W_i and the survival coefficient λ:

$$W_{i+1} = \lambda W_i.$$

Conversely, we can use this formula to determine the survival coefficient λ_i, using the ratio of the succeeding value of the weight to the preceding one:

$$\lambda = \frac{W_{i+1}}{W_i}.$$

It is this formula that we will use, but first we make one more remark. When modelling the trajectories we can "perform" the annihilation of a particle at each successive collision. However, we can proceed as follows: we can "perform" the annihilation of a particle only at "true" collisions, while the particle obviously survives in delta scattering but its weight is multiplied by the survival coefficient (of course, this weight then enters as a cofactor into the contribution function). This computational method is well known and justified. Our task is to define how the weight of a particle changes, on the average, from a "true" collision to a succeeding one.

Let the weight of a particle equal W_i at a "true" collision point. We are to find the weight of the particle at a point succeeding the "true" collision. Using the same reasonings as those for the function $\widetilde{f}$, we find that this weight is equal to $q\lambda W_i$ with probability q, equal to $\lambda^2 W_i$ with probability qp, equal to $\lambda^3 W_i$ with probability qp^2, and so on. Thus, the mean value of this weight is

$$W_{i+1} = W_i(q\lambda + qp\lambda + qp^2\lambda + \dots) = W_i \frac{q\lambda}{1 - p\lambda}.$$

In the case of pure scattering this formula yields the previous value of the survival coefficient

$$\widetilde{\lambda} = \lambda \cdot \frac{q}{1 - p\lambda}.$$

If $\lambda < 1$, the new survival coefficient decreases, since

$$\frac{q}{1 - p\lambda} = \frac{q}{(1-p) + p(1-\lambda)} = \frac{q}{q + p(1-\lambda)} < \frac{q}{q} = 1.$$

Consequently, the lifetimes of particles become shorter, which reduces the time of modelling.

Since our arguments imply that the path length density does not change when the absorption is taken into account, the parameters $\widetilde{\sigma}$ and $\widetilde{\lambda}$ also remain unchanged. These arguments can be made rigorous. To do this, in the case of absorption we need to consider the substochastic density of the path length normalized to the value $\lambda < 1$ rather than to unity as in the case of pure scattering.

2.5. RADIATION TRANSFER IN BROKEN CLOUDS

In this section we consider the application of the results in Section 5 to the concrete calculations of radiation transfer in a plane layer of broken clouds.

We consider the physical problem. There is a stochastic cloud layer. A homogeneous monodirectional flux of solar radiation is incident on the upper surface of this layer. We are to determine the scattered radiation flux reflected in a rigorously specified direction.

The averaged transport equation for the plane layer of broken clouds is described in detail in Glazov and Titiv (1980). This equation has the form (1), where the functions f and h have the special form:

$$f(t) = \sum_{i=1,2} \lambda_i \sigma_i \exp(-\sigma_i t), \tag{2.18}$$

$$h(r, \omega) = h_l(|r - r_0|)\delta_{\omega_\odot}(\omega),$$

where

$$h_l(t) = \sum_{i=1,2} \mu_i \sigma_i \exp(-\sigma_i t).$$

Here the parameters μ and λ satisfy the conditions

$$\sigma_1,\ \sigma_2 > 0,$$

$$\mu_1,\ \mu_2 > 0, \quad \mu_1 + \mu_2 = 1, \tag{2.19}$$

$$\lambda_1,\ \lambda_2 > 0, \quad \lambda_1 + \lambda_2 = 1. \tag{2.20}$$

They depend on the direction vector ω, but we omit this argument in their notation for convenience because this does not lead to confusion.

2.5.1. Free path density in the layer of broken clouds

Both densities f and h are probabilistic combinations of exponential densities and they are convenient for modelling. We are to model a random number with uniform density on the interval $(0, 1)$ and then a random value according to one of these exponential densities depending on the interval in which this random number appears.

Now we use formulas (15) and (16) from Subsection 5.3 to calculate the modified densities of the free path. We consider the formulas for the function $\widetilde{f}$. We obtain the formula for $\widetilde{h}$ in a completely analogous way and therefore, we give only the finite expression for it.

However, before proceeding to the function $\widetilde{f}$ for the transport equation in broken clouds, we must see what expression is obtained in the standard case, in which the density of the free path has the form

$$f_\sigma(l) = \sigma(r(l)) \exp(-\tau(l)),$$

where

$$\tau(l) = \int_0^l \sigma(r(t))dt$$

is the optical free path. For simplicity, from now on let σ be constant. This limitation can be removed if necessary. The formulas obtained are used when considering the density (49). The Laplace transform of the function f_σ is

$$\widehat{f}_\sigma(s) = \frac{\sigma}{s+\sigma}.$$

We substitute this expression into (15):

$$\widehat{(\widetilde{f}_\sigma)}(s) = \frac{q\dfrac{\sigma}{s+\sigma}}{1 - p\dfrac{\sigma}{s+\sigma}} = \frac{q\sigma}{s+q\sigma}.$$

On the other hand, it is obvious that

$$\frac{q\sigma}{s+q\sigma} \equiv \widehat{f}_{q\sigma}(s).$$

Hence,

$$\widehat{(\widetilde{f}_\sigma)}(s) = \widehat{f}_{q\sigma}(s).$$

If we apply the inverse Laplace transform to this expression, we find that in the standard case the simple relation holds because of special form of the path length density f:

$$\widetilde{f}_\sigma = f_{q\sigma},$$

i. e. we obtain the modified path length density from the initial density, replacing the cross-section σ by $q\sigma$. This accounts for the simplicity of the transform (9) in the standard case. We now proceed to the density (18). The Laplace transform of the function f is

$$\widehat{f}(s) = \sum_{i=1,2} \lambda_i \frac{\sigma_i}{s+\sigma_i}.$$

If we substitute this expression into (15), we obtain

$$\widehat{(\widetilde{f})}(s) = \frac{q\sum \lambda_i \dfrac{\sigma_i}{s+\sigma_i}}{1-p\sum \lambda_i \dfrac{\sigma_i}{s+\sigma_i}},$$

$$\widehat{(\widetilde{f})}(s) = \frac{q(\lambda_1\sigma_1+\lambda_2\sigma_2)s + q\sigma_1\sigma_1}{s^2+((\lambda_2+q\lambda_1)\sigma_1+(\lambda_1+q\lambda_2)\sigma_2)s+q\sigma_1\sigma_2}.$$

For brevity, we designate

$$\begin{aligned} b &= (\lambda_2+q\lambda_1)\sigma_1+(\lambda_1+q\lambda_2)\sigma_2, \\ c &= q\sigma_1\sigma_2, \\ d &= q(\lambda_1\sigma_1+\lambda_2\sigma_2), \end{aligned}$$

and obtain

$$\widehat{(\widetilde{f})}(s) = \frac{ds+c}{s^2+bs+c}.$$

The form of the density $\widetilde{f}$ essentially depends on the relationship between the coefficients b, c and d, in this formula. In order to understand the form of this density, we must consider these coefficients in greater detail. To this end we prove several simple algebraic equalities and inequalities for these coefficients, taking into account that

$$\sigma_1,\ \sigma_2 > 0, \qquad \lambda_1,\ \lambda_2 > 0, \quad \lambda_1+\lambda_2 = 1.$$

Statement 1. *The discriminant D of the quadratic trinomial s^2+bs+c is a positive value:*

$$D \equiv b^2 - 4c > 0.$$

Proof. From the inequality for arithmetical mean and geometrical mean we have

$$q\sigma_1 + \sigma_2 \geq 2\sqrt{q\sigma_1\sigma_2},$$

$$\sigma_1 + q\sigma_2 \geq 2\sqrt{q\sigma_1\sigma_2},$$

and at least one of these inequalities is strict since the two equalities $q\sigma_1 = \sigma_2$ and $\sigma_1 = q\sigma_2$ cannot hold simultaneously if $\sigma_1, \sigma_2 > 0$ and $q \neq 1$.

Therefore, the following inequality is also strict:

$$\lambda_1(q\sigma_1 + \sigma_2) + \lambda_2(\sigma_1 + q\sigma_2) > \lambda_1 \cdot 2\sqrt{q\sigma_1\sigma_2} + \lambda_2 \cdot 2\sqrt{q\sigma_1\sigma_2} = 2\sqrt{q\sigma_1\sigma_2},$$

since $\lambda_1 + \lambda_2 = 1$. Transforming the left-hand side of this inequality, we get

$$(\lambda_2 + q\lambda_1)\sigma + (\lambda_1 + q\lambda_2)\sigma_2 > 2\sqrt{q\sigma_1\sigma_2},$$

$$b > 2\sqrt{q\sigma_1\sigma_2}, \qquad b^2 > 4c, \qquad D = b^2 - 4c > 0.$$

□

Thus, the discriminant of the polynomial s^2+bs+c is positive and hence, its roots are real numbers. We denote them by $-\widetilde{\sigma}_1$, $-\widetilde{\sigma}_2$:

$$\widetilde{\sigma}_1 = \frac{b+\sqrt{D}}{2}, \qquad \widetilde{\sigma}_2 = \frac{b-\sqrt{D}}{2}.$$

We can now rewrite the trinomial $s^2 + bs + c$ as

$$s^2 + bs + c = (s + \widetilde{\sigma}_1)(s + \widetilde{\sigma}_2),$$

hence, the function $\widehat{(\widetilde{f}_\sigma)}(s)$ has the form

$$\widehat{(\widetilde{f}_\sigma)}(s) = \frac{ds + c}{(s + \widetilde{\sigma}_1)(s + \widetilde{\sigma}_2)}.$$

Statement 2. *The new coefficients $\widetilde{\sigma}_1$ and $\widetilde{\sigma}_2$ satisfy the inequalities*

$$\widetilde{\sigma}_1 > \widetilde{\sigma}_2 > 0.$$

Proof. We have $c = q\sigma_1\sigma_2 > 0$ and therefore

$$b^2 > b^2 - 4c.$$

Both sides of this equation are positive (see the preceding statement). So we can extract the square root. Since $b > 0$, we have the following inequality:

$$b > \sqrt{b^2 - 4c}$$

which is equivalent to

$$\widetilde{\sigma}_2 = \frac{b - \sqrt{D}}{2} > 0,$$

and we have the first of the two desired inequalities: $\widetilde{\sigma}_2 > 0$. To prove the second one, we just note that

$$\sqrt{D} = \sqrt{b^2 - 4c} > 0$$

and, hence,

$$\widetilde{\sigma}_1 - \widetilde{\sigma}_2 = \frac{b + \sqrt{D}}{2} - \frac{b - \sqrt{D}}{2} = 2\sqrt{D} > 0.$$

This is equivalent to

$$\widetilde{\sigma}_1 > \widetilde{\sigma}_2,$$

which completes the proof. □

Taking into account the above inequality, the inverse Laplace transform of the function $\widehat{(\widetilde{f}_\sigma)}(s)$ is

$$\widetilde{f}(x) = \frac{d\widetilde{\sigma}_1 - c}{(\widetilde{\sigma}_1 - \widetilde{\sigma}_2)} \exp(-\widetilde{\sigma}_1)t + \frac{c - d\widetilde{\sigma}_2}{(\widetilde{\sigma}_1 - \widetilde{\sigma}_2)} \exp(-\widetilde{\sigma}_2)t.$$

We represent this expression as

$$\widetilde{f}(x) = \widetilde{\lambda}_1\widetilde{\sigma}_1 \exp(-\widetilde{\sigma}_1)t + \widetilde{\lambda}_2\widetilde{\sigma}_2 \exp(-\widetilde{\sigma}_2)t.$$

Here

$$\widetilde{\lambda}_1 = \frac{d\widetilde{\sigma}_1 - c}{(\widetilde{\sigma}_1 - \widetilde{\sigma}_2)\widetilde{\sigma}_1}, \qquad \widetilde{\lambda}_2 = \frac{c - d\widetilde{\sigma}_2}{(\widetilde{\sigma}_1 - \widetilde{\sigma}_2)\widetilde{\sigma}_2}.$$

Statement 3. *The coefficients $\widetilde{\lambda}_1$ and $\widetilde{\lambda}_2$ satisfy the relations*

1) $\widetilde{\lambda}_1 + \widetilde{\lambda}_2 = 1,$

2) $\widetilde{\lambda}_1, \widetilde{\lambda}_2 > 0.$

Proof.

1) This assertion follows directly from the definition of the coefficients $\widetilde{\lambda}_1$ and $\widetilde{\lambda}_2$. We have

$$\frac{d\widetilde{\sigma}_1 - c}{(\widetilde{\sigma}_1 - \widetilde{\sigma}_2)\widetilde{\sigma}_1} + \frac{c - d\widetilde{\sigma}_2}{(\widetilde{\sigma}_1 - \widetilde{\sigma}_2)\widetilde{\sigma}_2} = \frac{d\widetilde{\sigma}_1\widetilde{\sigma}_2 - c\widetilde{\sigma}_2 + c\widetilde{\sigma}_1 - d\widetilde{\sigma}_1\widetilde{\sigma}_2}{(\widetilde{\sigma}_1 - \widetilde{\sigma}_2)\widetilde{\sigma}_1\widetilde{\sigma}_2}.$$

Now, using the relation $c = q\sigma_1\sigma_2$, we have:

$$\frac{d\widetilde{\sigma}_1\widetilde{\sigma}_2 - c\widetilde{\sigma}_2 + c\widetilde{\sigma}_1 - d\widetilde{\sigma}_1\widetilde{\sigma}_2}{(\widetilde{\sigma}_1 - \widetilde{\sigma}_2)\widetilde{\sigma}_1\widetilde{\sigma}_2} = \frac{q\sigma_1\sigma_2}{\widetilde{\sigma}_1\widetilde{\sigma}_2} = \frac{4q\sigma_1\sigma_2}{b^2 - (b^2 - 4c)} = \frac{4q\sigma_1\sigma_2}{4q\sigma_1\sigma_2} = 1.$$

2) First transform the inequality $\widetilde{\lambda}_1 > 0$. We have just proved that $\widetilde{\sigma}_1 > \widetilde{\sigma}_2 > 0$, therefore, the inequality

$$\frac{d\widetilde{\sigma}_1 - c}{(\widetilde{\sigma}_1 - \widetilde{\sigma}_2)\widetilde{\sigma}_1} > 0$$

is equivalent to

$$d\widetilde{\sigma}_1 > c.$$

Here we substitute the expression for $\widetilde{\sigma}_1$ and obtain

$$\frac{b + \sqrt{D}}{2} > \frac{c}{d}.$$

Then we multiply the numerator and the denominator in the left-hand side of this inequality by $b - \sqrt{D}$ and obtain

$$\frac{2c}{b - \sqrt{D}} > \frac{c}{d}$$

or

$$d > \frac{b - \sqrt{D}}{2}.$$

Now we transform this inequality the way we transformed the inequality $\widetilde{\lambda}_1 > 0$ and obtain

$$\frac{c - d\widetilde{\sigma}_2}{(\widetilde{\sigma}_1 - \widetilde{\sigma}_2)\widetilde{\sigma}_2} > 0, \qquad c > d\widetilde{\sigma}_2, \qquad \frac{c}{d} > \frac{b - \sqrt{D}}{2},$$

$$\frac{c}{d} > \frac{2c}{b + \sqrt{D}}, \qquad \frac{b + \sqrt{D}}{2} > d.$$

So, we have proved the double inequality

$$\frac{b+\sqrt{D}}{2} > d > \frac{b-\sqrt{D}}{2},$$

which is equivalent to the double inequality

$$\sqrt{D} > 2d - b > -\sqrt{D}$$

or

$$\sqrt{D} > |2d - b|.$$

If we square both sides of the last inequality, we find that

$$D > (2d-b)^2, \qquad b^2 - 4c > b^2 - 4bd + 4d^2, \qquad bd > cb + d^2.$$

Rewrite the expression (23) for b in the following form:

$$b = q(\lambda_1\sigma_1 + \lambda_2\sigma_2) + (\lambda_2\sigma_1 + \lambda_1\sigma_2).$$

Substitute the expressions for b, c and d in the last inequality and write down a sequence of equivalent inequalities:

$$(q(\lambda_1\sigma_1+\lambda_2\sigma_2)+(\lambda_2\sigma_1+\lambda_1\sigma_2))\cdot q(\lambda_1\sigma_1+\lambda_2\sigma_2) > q\sigma_1\sigma_2+q^2(\lambda_1\sigma_1+\lambda_2\sigma_2)^2,$$

$$q^2(\lambda_1\sigma_1+\lambda_2\sigma_2)^2+q(\lambda_1\sigma_1+\lambda_2\sigma_2)(\lambda_2\sigma_1+\lambda_1\sigma_2) > q\sigma_1\sigma_2+q^2(\lambda_1\sigma_1+\lambda_2\sigma_2)^2,$$

$$q(\lambda_1\sigma_1 + \lambda_2\sigma_2)(\lambda_2\sigma_1 + \lambda_1\sigma_2) > q\sigma_1\sigma_2,$$

$$(\lambda_1\sigma_1 + \lambda_2\sigma_2)(\lambda_2\sigma_1 + \lambda_1\sigma_2) > \sigma_1\sigma_2,$$

$$(\lambda_1^2 + \lambda_2^2)\sigma_1\sigma_2 + \lambda_1\lambda_2(\sigma_1^2 + \sigma_2^2) > \sigma_1\sigma_2.$$

Now use the inequality

$$\sigma_1^2 + \sigma_2^2 > 2\sigma_1\sigma_2,$$

which is strict since $\sigma_1 \neq \sigma_2$. Hence,

$$\begin{aligned}(\lambda_1^2 + \lambda_2^2)\sigma_1\sigma_2 + \lambda_2\lambda_2(\sigma_1^2 + \sigma_2^2) &> (\lambda_1^2 + \lambda_2^2)\sigma_1\sigma_2 + 2\lambda_1\lambda_2\sigma_1\sigma_2 \\ &= (\lambda_1^2 + \lambda_2^2 + 2\lambda_1\lambda_2)\sigma_1\sigma_2 = (\lambda_1+\lambda_2)^2\sigma_1\sigma_2 = \sigma_1\sigma_2.\end{aligned}$$

This completes the proof of the assertion. □

Propositions 5.2 and 5.3 show that the new coefficients $\widetilde{\sigma}_1$, $\widetilde{\sigma}_2$, $\widetilde{\lambda}_1$ and $\widetilde{\lambda}_2$ satisfy the same conditions (19) and (20) as the initial coefficients σ_1, σ_2,

λ_1 and λ_2. Thus, we again represent the new path length density $\widetilde{f}$ as a mixture of exponential densities:

$$\widetilde{f}(t) = \sum_{i=1,2} \widetilde{\lambda}_i \widetilde{\sigma}_i \exp(-\widetilde{\sigma}_i t).$$

The modelling algorithm remains unchanged and therefore, we can use the above formula for modelling the path length, replacing only the parameters σ_1, σ_2, λ_1, and λ_2 by $\widetilde{\sigma}_1$, $\widetilde{\sigma}_2$, $\widetilde{\lambda}_1$, and $\widetilde{\lambda}_2$.

2.5.2. Results of numerical experiments

In order to check the efficiency of the third method, we carried out a numerical experiment to calculate the intensity of the solar radiation flux reflected from a stochastic cloud layer.

According to the algorithm in Subsection 5.3 and the formulas in Subsection 6.1, we first choose some ε $(0 < \varepsilon \ll 1)$. We divide the interval $(-1, 1)$ into two parts: $(-1, \varepsilon)$ and $(\varepsilon, 1)$. Then we calculate the quantities

$$p(\varepsilon) = \int_{\varepsilon}^{1} g(\mu) d\mu,$$

$$q(\varepsilon) = 1 - p(\varepsilon)$$

and determine the new phase function $\widetilde{g}$ and the path length density $\widetilde{f}$.

In the calculations we varied the value of the survival coefficient $\widetilde{\lambda}$ of a particle at a collision and the value ε. We choose the value ε according to the requirements for the calculation accuracy and the number of trajectories modelled. An increase or decrease in ε affects the estimate of the flux by the Monte Carlo method. It is not difficult to understand that the quantity $\sqrt{2\varepsilon}$ is, in fact, the upper bound of the value of the angle (in radians) through which a particle is deflected from the true direction with probability p at a collision. Note that if α $(\alpha > 0)$ is a deflection angle, then

$$\alpha \sim \sin\alpha = \sqrt{1 - \cos^2\alpha} \leq \sqrt{1 - (1-\varepsilon)^2} \sim \sqrt{2\varepsilon}.$$

Therefore, an excessive increase in ε results in an increase in the bias of the estimate. When the values ε are too small, the function $\widetilde{g}$ involves the larger values of the highly peaked component of the initial phase function g, which leads to an increase in the variance of the estimate and greatly reduces its efficiency. Unfortunately, we cannot give the necessary recommendation as to the choice of the optimal value ε. This value strongly depends on the form of a particular phase function g. We should probably choose ε in accordance with the concrete requirements for the computational result and using the graph of the phase function g. As to the relationship between the calculation error (more precisely, the bias of the estimate) and the value of the survival coefficient λ, we can say that as the value of λ decreases, the lifetime of a particle becomes shorter on the average. Consequently, the direction distortions of sections of the trajectory which result from the substitution $g \to g_{tr}$ become smaller. Besides, the displacement decreases, the computational results become more accurate, while the number of the trajectories modelled remains unchanged.

As mentioned above, the scattering phase function in this problem is a function of a single variable. The initial phase function g is specified at the nodes of the segment $[-1, +1]$ and is interpolated linearly between the nodes. The values of the phase function g_i at the nodes $-1 = \mu_{60} < \mu_{59} < \ldots < \mu_1 = +1$ are given in Table 1.

The graph of this phase function on a logarithmic scale is shown in Figure 4a.

Table 2 shows how the integral of the phase function

$$\int_{\mu}^{1} g(\mu) d\mu$$

changes as $\mu \to 1$.

Table 3 gives the computational results for different values of the "threshold" quantity ε (the survival coefficient is equal to unity here) and the comparison results of the efficiency (computational cost) of the calculation for the modified and nonmodified computational methods. With the optimal choice of ε the error of the flux approximation was less than 0.1% (three significant digits of the result coincide with the test ones), and the computation time of the modified scheme is 40 times less than that of the nonmodified scheme.

Table 4 gives the relationship between the relative computation error due to the displacement and the survival coefficient λ. The value ε is deliberately taken to be nonoptimal.

Table 1: Values of the scattering phase function at nodes

μ_i	g_i	μ_i	g_i
1.0000000	880.825000	0.3420202	0.071725
0.9999996	879.977000	0.2588190	0.050475
0.9999985	877.439000	0.1736482	0.036268
0.9999966	873.227000	0.0871557	0.026828
0.9999939	867.370000	0.0000000	0.020637
0.9999905	859.905000	-0.0871557	0.017039
0.9999619	800.485000	-0.1736482	0.014835
0.9999143	711.633000	-0.2588190	0.013781
0.9998477	605.686000	-0.3420202	0.014595
0.9993908	223.641000	-0.4226183	0.018216
0.9986295	64.488100	-0.5000000	0.022489
0.9975641	23.896300	-0.5735765	0.025527
0.9961947	13.204200	-0.6427876	0.032559
0.9945219	9.106590	-0.7071068	0.060887
0.9925461	6.998930	-0.7660444	0.117805
0.9902681	5.745800	-0.8660254	0.080565
0.9876884	4.930390	-0.9063078	0.076471
0.9848077	4.365700	-0.9396926	0.075500
0.9659258	2.917150	-0.9659258	0.076960
0.9396926	2.129210	-0.9848077	0.091419
0.9271839	1.885840	-0.9876884	0.096261
0.9063078	1.560290	-0.9902681	0.101405
0.8660254	1.133690	-0.9925461	0.111265
0.8191521	0.817229	-0.9945219	0.130723
0.7660444	0.585151	-0.9961947	0.163201
0.7071068	0.417133	-0.9975641	0.211489
0.6427876	0.295835	-0.9986295	0.249238
0.5735765	0.208708	-0.9993908	0.208409
0.5000000	0.146255	-0.9998477	0.219326
0.4226183	0.102412	-1.0000000	0.330932

Table 2: Values of the integral of the phase function

μ_i	s_i	μ_i	s_i
1.0000000	0.0000000	0.3420202	0.9473060
0.9999996	0.0003199	0.2588190	0.9521580
0.9999985	0.0012779	0.1736482	0.9556830
0.9999966	0.0028683	0.0871557	0.9582880
0.9999939	0.0050822	0.0000000	0.9602620
0.9999905	0.0079067	-0.0871557	0.9618290
0.9999619	0.0305334	-0.1736482	0.9631440
0.9999143	0.0648764	-0.2588190	0.9643070
0.9998477	0.1067620	-0.3420202	0.9654340
0.9993908	0.2875640	-0.4226183	0.9666960
0.9986295	0.3922350	-0.5000000	0.9681990
0.9975641	0.4371720	-0.5735765	0.9698850
0.9961947	0.4614150	-0.6427876	0.9718030
0.9945219	0.4792250	-0.7071068	0.9746710
0.9925461	0.4944090	-0.7660444	0.9796970
0.9902681	0.5082630	-0.8660254	0.9891610
0.9876884	0.5214060	-0.9063078	0.9921790
0.9848077	0.5341840	-0.9396926	0.9946010
0.9659258	0.5998040	-0.9659258	0.9965090
0.9396926	0.6629740	-0.9848077	0.9980260
0.9271839	0.6869400	-0.9876884	0.9982840
0.9063078	0.7212700	-0.9902681	0.9985270
0.8660254	0.7730540	-0.9925461	0.9987590
0.8191521	0.8166910	-0.9945219	0.9989870
0.7660444	0.8522300	-0.9961947	0.9992210
0.7071068	0.8804180	-0.9975641	0.9994660
0.6427876	0.9023010	-0.9986295	0.9997010
0.5735765	0.9189640	-0.9993908	0.9998670
0.5000000	0.9314270	-0.9998477	0.9999600
0.4226183	0.9406090	-1.0000000	1.0000000

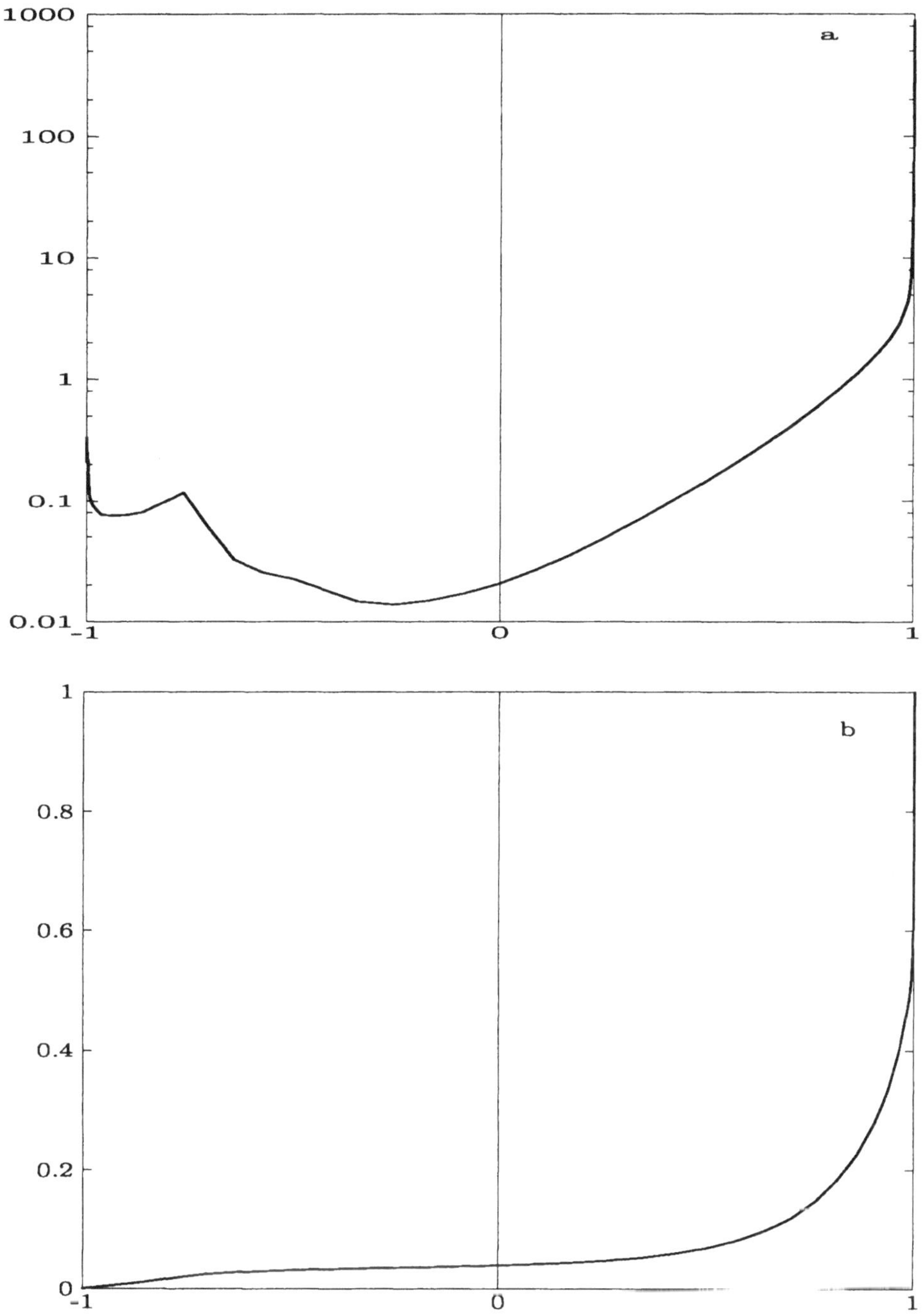

Figure 4: Graphs of the phase function on a logarithmic scale (a) and the integral of the phase function (b)

2.6. CONCLUSION

The results obtained can be used for solving three-dimensional problems and for the case in which the phase function depends on space points.

The use of the modifications proposed does not imply that other optimizing procedures cannot be used simultaneously (in particular, parallelizing the computational algorithm).

The result in Subsection 4.3 clearly shows that the information about the angular scattering of the reflected and transmitted radiation does not allow us to directly reconstruct the highly peaked scattering phase function. We are led to pose this problem differently, for example, we specify the "threshold" value ε and try to determine the regular part $\widetilde{g}$ of the phase function g. As to the highly peaked component, we should note that for the real physical phase function its highly peaked and regular components are not independent parts at least because the phase function has no discontinuity at their connection point. Therefore, in order to estimate the highly peaked component, we can use the corresponding statistical information and seek the phase function in a statistical ensemble of functions, but this is beyond the scope of this work.

Table 3: Error and computation time versus ε

value ε	relative error, %	computat. time, sec	value, ε	relative error, %	computat. time, sec
0.003805	5	31	0.012312	1	23
0.005478	3	29	0.015192	2	20
0.007454	2	27	0.034074	2	19
0.009732	0	26			

Table 4: Computation time versus λ

value λ	relative error, %	value λ	relative error, %
1.00	5.2	0.70	2.0
0.90	5.0	0.60	1.2
0.80	4.1	0.50	0.0

Chapter 3.

Calculation of the correlation characteristics of radiation field in the stochastic medium by the Monte Carlo method

Consider a process of radiation transfer in a domain R of 3-dimensional space, where the extinction coefficient $\sigma(r)$ is random ($\sigma(r) = 0$ outside R).

Then the functionals I of collision density are also random variables. Designate the mathematical expectation of I over random σ by angular brackets:

$$M_\sigma I = \langle I \rangle.$$

In this chapter we consider a problem of calculation of a correlation coefficient for random functionals I_1 and I_2:

$$B_{12} = B(I_1, I_2).$$

3.1. PROBLEM STATEMENT

For functionals I_k we may take, for instance, the values of intensity in points of the phase space. Using the accepted notation we may write it as follows:

$$B_{12} = \langle (I_1 - \langle I_1 \rangle)(I_2 - \langle I_2 \rangle) \rangle.$$

For simplicity let $\sigma(r)$ be represented in the form of a sum:

$$\sigma(r) = \overline{\sigma}(r) + \widetilde{\sigma}(r),$$

where

$$\overline{\sigma}(r) = \langle \sigma(r) \rangle,$$

$$\min_R \overline{\sigma}(r) = \sigma_0 > 0,$$

and the random field $\widetilde{\sigma}(r)$ admits finite decomposition:

$$\widetilde{\sigma}(r) = \sum_{i=0}^{N} p_i \xi_i \varphi_i(r), \tag{3.1}$$

where ξ_i are random independent values with zero mathematical expectation and dispersion (1), coefficients $p_i > 0$ and $\varphi_i(r)$ are nonrandom (deterministic) basic functions.

We also assume that $\varphi_0(r) \equiv 0$. In particular, if σ is a stationary field with a finite spectrum then $\widetilde{\sigma}$ may be represented as a finite sum

$$\widetilde{\sigma}(r) = \sum_{i=0}^{N} p_i(\xi_i \cos \lambda_i r + \eta_i \sin \lambda_i r),$$

where ξ_i, η_i are independent standard Gaussian random variables ($\lambda_0 = 0$).

We also assume that the numbers p_i are rather small, so that the probability of the event $\sigma(r) > 0$ is close to 1. We designate $q_i = p_i \xi_i$ and define the function F by

$$F_k(q_0, q_1, \dots, q_N) = I_k\left(\overline{\sigma} + \sum_{i=0}^{N} q_i \varphi_i\right) \equiv I_k(\sigma).$$

Designate its derivatives by

$$D_i^k = \frac{\partial F_k}{\partial q_i}(0, 0, \dots, 0), \quad i = 1, 2, \dots, N.$$

As the variation of the field $\sigma(r)$ is small, we may use a linear approximation for the functionals I_k:

$$I_k(\sigma) \approx I_k(\overline{\sigma}) + \sum_{i=0}^{N} D_i^k q_i = I_k(\overline{\sigma}) + \sum_{i=0}^{N} D_i^k p_i \xi_i.$$

Hence,

$$\langle I_k \rangle \approx I_k(\overline{\sigma}).$$

For a correlation moment B_{12} we have

$$B_{12} \approx \left(\sum D_i^1 p_i \xi_i, D_i^2 p_j \xi_j\right).$$

Hence,

$$B_{12} \approx \sum_{i=0}^{N} p_i^2 D_i^1 D_i^2. \tag{3.2}$$

The derivatives D_i^k can be computed (see Section 00002) by the Monte Carlo method using the same trajectories as the main functionals I_k.

Consider the formula (2).

Assume that the decomposition (1) for the random component $\widetilde{\sigma}$ of the field $\sigma(r)$ consists of only one item $p_0\xi_0$ $(\varphi_0 \equiv 1)$ and, besides, $\overline{\sigma}(r) \equiv \overline{\sigma} =$ const. It is clear that in the accepted notation

$$B_\sigma \equiv \langle(\sigma - \overline{\sigma})^2\rangle = p_0^2.$$

Let

$$D_0^k = \frac{\partial I_k}{\partial \overline{\sigma}}.$$

Then (2) takes the form

$$B_{12} \approx b_\sigma \frac{\partial I_1}{\partial \overline{\sigma}} \frac{\partial I_2}{\partial \overline{\sigma}}. \tag{3.3}$$

We cannot use this formula for the correlation moment B_{12}. If the decomposition (1) includes more than one item (that is, we take into account the random inhomogeneity of the field $\sigma(r)$), the other items essentially influence the value B_{12} and the formula (3) becomes incorrect. It only allows to evaluate dispersions of the functionals I_k in a random inhomogeneous medium.

An algorithm for calculation the correlation characteristics of a radiation field has been proposed in Antyufeev and Mikhailov (1976).

3.2. COMPUTATION OF DERIVATIVES OF A FIELD OF BRIGHTNESS BY THE MONTE CARLO METHOD

In this section we consider a problem of computing derivatives D_i of radiation field by the Monte Carlo method.

We consider a plane, horizontally homogeneous slab $\{0 \le z \le 1\}$ with a random scattering coefficient,

$$\sigma(r) \equiv \sigma(z), \quad \overline{\sigma}(r) \equiv \overline{\sigma} = \text{const}\,.$$

Estimate a flux of radiation leaving the slab through the boundary $z = 1$ in the direction ω^*. The functional I can be represented as a series:

$$I = (f, \varphi^*) = \sum_{n=0}^{\infty} (K^n \psi, \varphi^*). \tag{3.4}$$

Here

$f(x)$ is the collision density for particles in the phase space;

$x = (z, \omega)$ denotes generic point of the phase space;

z is the height of a particle (which is a geometric space coordinate);

$\omega = (a, b, c)$ denotes a unit direction vector of a particle just before collision, $a^2 + b^2 + c^2 = 1$.

Further we also use the following notation:

$$\begin{aligned} x_i &= (z_i, \omega_i), \\ \omega_i &= (a_i, b_i, c_i), \\ \omega^* &= (a^*, b^*, c^*). \end{aligned}$$

The collision density f satisfies the transport equation

$$f = Kf + \psi,$$

where K is an integral operator with kernel

$$k(x' \to x) = \frac{g(\omega' \cdot \omega)}{2\pi} \frac{\sigma(z)}{c} \exp\left[-\int_{z'}^{z} \frac{\sigma(t)}{c} dt \right],$$

ψ is a density function of radiation source. The contribution function of the functional I is

$$\varphi^*(x) = \frac{g(\omega, \omega^*)}{2\pi c^*} \exp\left[-\int_{z}^{1} \frac{\sigma(t)}{c^*} dt \right].$$

Substituting the expression for $\sigma(z)$ into the formulas, we obtain

$$\sigma(z) = \overline{\sigma} + \sum q_i \varphi_i(z).$$

Here $\varphi_0(z) \equiv 1$, $\varphi_i(z) = \cos(\lambda_i z)$ or $\sin(\lambda_i z)$ for $i \neq 0$.

Differentiate the n-th item in the sum (4) over the variable q_j (validity of the differentiation may be justified the way it has been done in Mikhailov (1967)):

$$\frac{\partial}{\partial q_j}\left[\int \psi(x_0)dx_0 \dots dx_n \prod_{i=1}^{n} k(x_{i-1} \to x_i)\varphi^*(x_n)\right]\Bigg|_{q=0}$$
$$= \int \psi \; dx_0 \dots dx_n \prod \varphi^* \left[\sum_{i=1}^{n} \frac{D_j k(x_{i-1} \to x_i)}{k(x_{i-1} \to x_i)} + \frac{D_j \varphi^*(x_n)}{\varphi^*(x_n)}\right]\Bigg|_{q=0}.$$

Denote by S_n the expression in square brackets in the last formula. To estimate the derivative $D_j I$ by the Monte Carlo method, we should model photon trajectories and bring the contribution $\varphi^*(x_n)S_n$ to the estimator.

We have:
for $j = 0$, $\lambda_0 = 0$, $\varphi_0 \equiv 1$:

$$S_n = \sum_{i=1}^{n} \left(\frac{1}{\overline{\sigma}} - \frac{z_i - z_{i-1}}{c_i}\right) - \frac{1 - z_n}{c^*};$$

for $j \neq 0$, $\varphi_j(z) = \sin(\lambda z)$:

$$S_n = \sum_{i=1}^{n} \left(\frac{\sin(\lambda z_i)}{\overline{\sigma}} + \frac{\cos(\lambda z_i) - \cos(\lambda z_{i-1})}{\lambda c_i}\right) + \frac{\cos\lambda - \cos(\lambda z_n)}{\lambda c^*};$$

for $j \neq 0$, $\varphi_j(z) - \cos(\lambda z)$:

$$S_n = \sum_{i=1}^{n} \left(\frac{\cos(\lambda z_i)}{\overline{\sigma}} + \frac{\sin(\lambda z_i) - \sin(\lambda z_{i-1})}{\lambda c_i}\right) + \frac{\sin\lambda - \sin(\lambda z_n)}{\lambda c^*}.$$

3.3. DISPERSIONS OF THE DERIVATIVES $D_j I$ ESTIMATOR

In this section we prove that dispersion of the derivatives $D_j I$ estimator is limited.

Denote for brevity

$$d(x) = \int dy\ k(x,y) \le 1, \qquad \widetilde{k}(x,y) = \frac{k(x,y)}{d(x)},$$

$$\zeta_n = \varphi^*(x_n) S_n \prod_{j=1}^{n} d(x_j).$$

Write down the n-th item of the correspondent series for $D_j I$. We have

$$\int \psi(x_0)\, dx_0 \dots dx_n \prod_{i=1}^{n} k(x_{i-1}, x_i) \varphi^*(x_n) S_n$$
$$= \int \psi(x_0)\, dx_0 \dots dx_n \prod_{i=1}^{n} \widetilde{k}(x_{i-1}, x_i) \left[\varphi^*(x_n) S_n \prod_{j=1}^{n} d(x_j) \right] = M\zeta_n,$$

where mathematical expectation is taken over the joint density for the distribution of the points $x_0, x_1, \dots, x_n$.

We have the following estimate for $|\zeta_n|$:

$$|\zeta_n| \le C \left(\sum_{i=1} \right)^n \frac{1}{|c_i|} \prod_{j=1}^{n} d(x_j) \varphi^*(x_n).$$

Hence,

$$M\zeta_n^2 \le C^2 n M \left[\sum_{i=1}^{n} \frac{1}{c_i^2} \prod_{j=1}^{n} d(x_j) \varphi^*(x_n) \right]$$
$$\le C^2 n \int \psi(x_0)\, dx_0 \dots dx_n \prod_{i=1}^{n} k(x_{i-1}, x_i) \sum_{j=1}^{n} \frac{1}{c_j^2} \varphi^*(x_n)$$
$$= C^2 n \sum_{j=1}^{n} \int \psi(x_0)\, dx_0 \dots dx_n \cdot k(x_0, x_1) \dots k(x_{j-2}, x_{j-1})$$
$$\times \frac{k(x_{j-1}, x_j)}{c_j^2} \cdot k(x_j, x_{j+1}) \dots k(x_{n-1}, x_n) \cdot \varphi^*(x_n).$$

Let $\overline{K}$ be an integral operator with kernel

$$\overline{k}(x,y) = \frac{k(x,y)}{c_y^2},$$

where $y = (z_y, \omega_y)$, $\omega_y = (a_y, b_y, c_y)$. It is clear that k is a limited function and $\|K\| = p < \infty$, and $\|K^2\| = q < 1$.

Therefore,

$$\|K^{j-1}\overline{K}K^{n-j}\| \leq q^{(j-1)/2} p q^{(n-j)/2} \leq \frac{p}{q^2} q^{n/2}.$$

Hence,

$$M\zeta_n^2 \leq C^2 n^2 \frac{p}{q^2} q^{n/2} \|\psi\| \|\varphi^*\| = C_1 n^2 q^{n/2}.$$

Here the constant C_1 does not depend on n.

Now, for the dispersion of the sum $\sum_{n=1}^{m} \zeta_n$ we have the following estimate:

$$\left[D\left(\int \zeta_n\right)\right]^{1/2} \leq \sum (D(\zeta_n))^{1/2} \leq C_1^{1/2} \sum n q^{n/4}.$$

Here the series $\sum n q^{1/4}$ converges, which completes the proof.

Chapter 4.

Identification of the phase function

Phase function is one of the most important parameters of the Earth's atmosphere. Solving each problem in radiation transfer, one needs to know this function. In this section we will study the problem of estimating the phase function, using results of measurements of the sky brightness.

Not only single-scattered radiation causes the brightness of the sky but multiscattered radiation and the radiation reflected from the earth's (bottom) surface as well. Thus the angular distribution of brightness over the sky only to a certain extent corresponds to a form of the natural phase function.

Many researchers have been solving the problem of separation of the first order phase function from the multiscattered radiation indicatrix for a long time by different means based on special approximate relations. However, many obtained solutions do not have satisfactory accuracy and cannot be used in many other actual physical problems.

Theoretical calculations have shown that we cannot put the equality sign between the actual phase function (indicatrix of the first order) and the indicatrix of multiscattered radiation, even in the case of small optical thickness of the atmosphere when the share of multiscattered radiation is relatively small. This resulted in attempts to separate the scattering indicatrix from the brightness indicatrix. To separate the first order indicatrix the authors usually apply relations they have found from natural experiments. These relations allow to perform the separation of intensity components of single-scattered and multiscattered radiation for at least one scattering angle. After that, they set the multiscattered radiation component constant over all the scattering angles (that is, they regarded the multiscattered radiation indicatrix as a spherical one). Then the scattering indicatrix of the first order can be found by simple subtraction of this constant value from the whole indicatrix. This way leads to significant errors because the actual indicatrix of multiscattered radiation has a form that is far from spherical one. The most significant errors appear at large scattering angles.

From the preceding it follows that, to separate the phase function, we need to obtain precise values of the multiscattered component and its angular distribution.

For this reason some authors use data from the calculations that have been performed for anisotropic scattering (see Feygelson, Malkevich, *et. al.* (1962) and Atroshenko, Glazova, *et. al.* (1962)). However, in that calculation they have used some particular phase functions being in specific conditions. They cannot be used rigorously in other conditions.

4.1. STATEMENT OF THE PROBLEM

The only way to eliminate these problems is to apply a method of successive approximations. In our study we will use the following notation.

ω: observation direction;

θ: zenith angle of the observation direction;

$\omega_\odot$: direction opposite to the solar radiation direction;

$\theta_\odot$: zenith solar angle;

τ: optical thickness of the atmosphere in the observation direction;

Γ: coefficient of asymmetry of the indicatrix $g(\mu)$, which is defined as the ratio of the fluxes scattered to the forward and backward semispheres

$$\Gamma = \frac{\int_0^1 g(\mu)d\mu}{\int_{-1}^0 g(\mu)\mu d\mu}.$$

The indicatrix of brightness gives a characteristic of total scattered radiation, which includes molecular (Rayleigth), aerosole once-scattered and multiscattered radiation. It can be presented as a sum

$$g_b(\mu) = g(\mu) + g_m(\mu),$$

and in turn

$$g(\mu) = g_a(\mu) + g_R(\mu).$$

The main idea of the proposed method is to separate single-scattered radiation intensity J, which, in the case of plane atmosphere model, is related to the phase function by a simple formula

$$J(\omega) = \frac{\exp(-\tau)}{2\pi \cos\theta} \cdot g(\omega \cdot \omega_{\odot}).$$

Further we call the angle between the observation direction ω and the solar direction $\omega_{\odot}$ a single-scattering direction. If the values I_ω, τ, θ, and the single-scattering angle are known, then the value $g(\mu)$ $(\mu = \omega \cdot \omega_{\odot})$ can be easily found from this relation.

Recall that we use a piecewise approximation for the phase function $g(\mu)$. Thus, the phase function can be specified by a vector $g = (g_1, \ldots, g_n)$ with the components $g_k = g(\mu_k)$.

To find $g_k = g(\mu_k)$, we need to know $J(\omega_k)$, where ω_k is chosen so that

$$\omega_k \cdot \omega_{\odot} = \mu_k.$$

Besides, it is necessary to know the value τ_k of optical thickness of the atmosphere in the observation direction ω_k.

Now choose the following observation scheme (see Figure 5). The radiation detector is located on the earth's surface and measures intensities of scattered solar radiation in several directions that make up equal zenith angles $\theta = \theta_{\odot}$ ($\theta_{\odot}$ is the solar zenith angle) and different azimuth angles φ (φ is the angle between the vertical plane of observation direction and the plane of the solar vertical, $0° < \varphi < 180°$).

In this scheme the angle of single scattering can range from zero to $2\theta < 180°$ (here the angle of single scattering is equal to the angle between the observation direction and the direction toward the sun). All the other optical parameters of the atmosphere model, namely the coefficients of molecular and aerosol scattering and absorption, the albedo, and the scattering indicatrix of the underlying surface, are assumed to be known.

The problem is of estimating the scattering indicatrix using the measured values $I_1^*, \ldots, I_m^*$. The scattering indicatrix is determined completely by its values $g_1, g_2, \ldots g_n$ at the nodes and therefore the problem amounts to determining the values $g_1, g_2, \ldots g_n$. Note that due to the problem geometry (viz. $2\theta < 180°$) the number m of observations is less than the number n of unknown values.

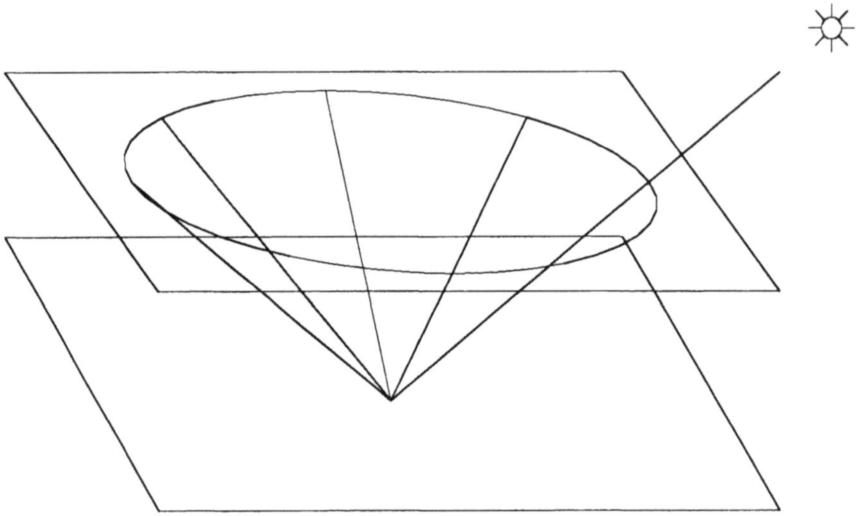

Figure 5: Observation scheme

The indicatrix is recovered by using the iterative algorithm. The algorithm is based on the separation of single-scattered radiation from the total flux of observed scattered solar radiation.

Denote by $I_1, \ldots, I_m$ the intensities observed at the detector point:

$$I_k = I_k(g_1, \ldots, g_n).$$

In this expression I can be presented as a sum

$$I = J + R,$$

where J is a single-scattered component and R is a multiscattered one.

As we have noted earlier, for the model in hand the corresponding intensity of single-scattered radiation is determined by the simple formula

$$J_k = C \cdot g_k,$$

where

$$C = \frac{\exp(-\tau)}{2\pi} \cos\theta_{\odot},$$

and τ is the optical thickness of atmosphere in the direction of observation. Note that the thickness is the same for all the directions ω_k and therefore, the constant C also does not depend on the number of the corresponding direction. To construct an iterative method, use the following notice. Although the multiscattered component R is not isotropic, it is still less sensitive of the phase function than the once-scattered component. This notice makes up a background for two iterative methods — "additive" method and "multiplicative" method that are described below.

4.2. METHOD OF SUCCESSIVE APPROXIMATIONS

Both methods are based on separating the "main part" from the vector of observed values.

4.2.1. "Additive" method

Taking into account that $g \approx g^*$ and the notice in the end of the preceding section, we can set approximately

$$R_k \approx R_k^*,$$

where

$$R_k \equiv R_k(g_1, g_2, \dots g_n), \qquad R_k^* \equiv R_k^*(g_1^*, g_2^*, \dots g_n^*).$$

Hence, we have

$$I_k^* - I_k \approx J_k^* - J_k = C(g_k^* - g_k).$$

The values I_k can be found by any numerical method for solving the equation for radiation transport. In our computation we used the Monte Carlo method. The values I_k^* are obtained from the measurements described above.

Thus, we obtain an iteration process. At every step we have a simple diagonal system of linear equations for unknown δ_k

$$C\delta_k = I_k^* - I_k^{(k)}, \quad m = 1, \dots, n,$$

$$\delta_m = g_m^{(k)} - g_m^{(k-1)}, \qquad I_m^{(k)} = I_m(g_1^{(k)}, \dots, g_n^{(k)}),$$

$g_m^{(k)}$ is the k-th approximation to the m-th component of the sought vector $g^* = (g_1^*, \dots, g_n^*)$.

We can easily write down a precise formula to express the following iteration:

$$g_k^{(i+1)} = g_k^{(i)} + \frac{2\pi \cos\theta_\odot}{\exp(-\tau)}(I_k^* - I_k^{(i)}), \quad k = 1, \dots, n.$$

Since $r < n$, at each iteration we recalculate only r components of n. We call the rest $(n - r)$ components a "tail" of the phase function. To find it we take an "averaged" phase function, multiply its r to n components by such a factor that its the r-th component equals to r-th component of the estimated phase function. Then we put unknown components $g_r, \dots, g_n$ equal to the relevant components of the "averaged" phase function.

4.2.2. "Multiplicative" method

This time we set an approximate equality for proportions

$$\frac{J(g^*)}{I(g^*)} \approx \frac{J(g)}{I(g)}.$$

With this relation we arrive at the "multiplicative" iteration

$$g_i^{(k+1)} = \frac{I_i^* J_i(g^{(k)})}{2\pi \exp(\tau) I_i(g^{(k)})}$$

and the following norming of the obtained iteration g^{k+1}.

The comparative effectiveness of the "additive" and "multiplicative" methods generally depends on the optical conditions of the atmosphere.

4.3. ESTIMATING THE INTENSITY OF SCATTERED RADIATION

The iterative method proposed in Section 2 requires very accurate estimation of intensity of scattered radiation. The relative error of estimation must not exceed 1% to 2%. In this calculation we used a model atmosphere as a plane slab. Intensity computations for this model have been carried out by the Monte Carlo method.

On the suitability of the plane slab model see the next section.

4.3.1. On the use of the plane atmosphere model

A number of computations have been carried out to estimate the error of approximation of the spherical atmosphere model by the plane one in the considered problem. We used the "conjugate" scheme in computations for the spherical and the plane models. To make comparison more precise, we arranged the computer program so that the beginning of each pseudo-random trajectory in the plane model was the same as in the spherical model.

Table 1: Relative error of the plane atmosphere model approximation compared with the spherical one

φ°	Variance		φ°	Variance	
	a	b		a	b
0	-1.2	0.8	60	-0.7	1.5
1	-1.1	0.8	70	-0.8	2.0
2	-1.0	0.8	80	-0.6	1.2
4	-0.9	1.0	90	-0.3	1.4
6	-0.8	0.9	100	-0.3	1.5
8	-0.6	1.0	110	-0.3	1.7
10	-1.1	1.0	120	0.0	1.8
15	-0.7	1.1	130	0.3	1.8
20	-1.3	1.1	140	0.3	1.8
30	-0.7	1.1	150	0.3	1.8
40	-1.2	1.4	155	0.2	1.7
50	-0.6	2.1			

Comparison computations have been carried out for two different cases with the following parameter values:

a) $\theta_{\odot} = 80^{\circ}$, $\tau = 0.4$, $q = 0.1$, the phase function with $\Gamma = 2.8$;

b) $\theta_{\odot} = 80^{\circ}$, $\tau = 0.06$, $q = 0.2$, the phase function with $\Gamma = 9.3$.

Here q is the albedo of the earth's surface. Relative errors $\delta_p = (I_p - I_s)/I_s$ are presented in Table 1.

The chosen values of the parameters τ and q and the phase function correspond to the short-wave and long-wave parts of the studied specrum region. The role of multiscattering effect is quite different in those parts. It is also clear that spherical form of the atmosphere has the strongest bearing at large values of solar zenith angles. That is why we have taken $\theta_{\odot} = 80^{\circ}$, which is the largest value of zenith angle of the observation direction in the numerical experiments. Thus, Table 1 gives the limits of the error related to the choice of the plane atmosphere model. We can see that only in some cases the error reaches the 2% level.

4.3.2. "Local" estimate of intensity

In the Monte Carlo intensity computation for the plane atmosphere model we applied a local estimate. A few words about this estimate.

Let $I(r^*, \omega^*)$ be the needed value of intensity of the scattered radiation at a spatial point $r^* = (x^*, y^*, z^*)$ in a direction $\omega^* = (\omega_x^*, \omega_y^*, \omega_z^*)$. Then

$$I(r^*, \omega^*) = \int_R \int_\Omega \varphi(r, \omega) h^*(r, \omega) dr d\omega,$$

where $\varphi(r, \omega)$ is a collision density function defined in the phase space of points r and directions ω, $h^*(r, \omega)$ is a contribution function:

$$h^*(r, \omega) = \begin{cases} \dfrac{g(\mu^*)}{2\pi} \dfrac{\exp(-\tau)}{\omega_z}, & (z' - z)\omega_z^* > 0, \\ 0, & (z' - z)\omega_z^* < 0, \end{cases}$$

$$\mu^* = (\omega, \omega^*), \qquad \omega_z = \omega^* e_z, \qquad e_z = (0, 0, 1),$$

$$\tau = \int_0^{|r - r^*|} \sigma \left(r' + t \frac{r - r'}{|r - r'|} \right) dt$$

is the optical length of the segment $[r, r']$.

Let $\{r_n, \omega_n\}$ be a random chain of photon collisions we model on computer. Then

$$I(r^*, \omega) = M\xi,$$

where

$$\xi = \sum_{n=1}^{N} h^*(r_n, \omega_n).$$

Here N is the number of the trajectory termination as the photon leaves the atmosphere or is absorbed by the earth's surface.

Note that, using the plane atmosphere model, we can apply the Monte Carlo modification called "the method of mathematical expectation" and spare computation time. This development is described in detail in Chapter 1.

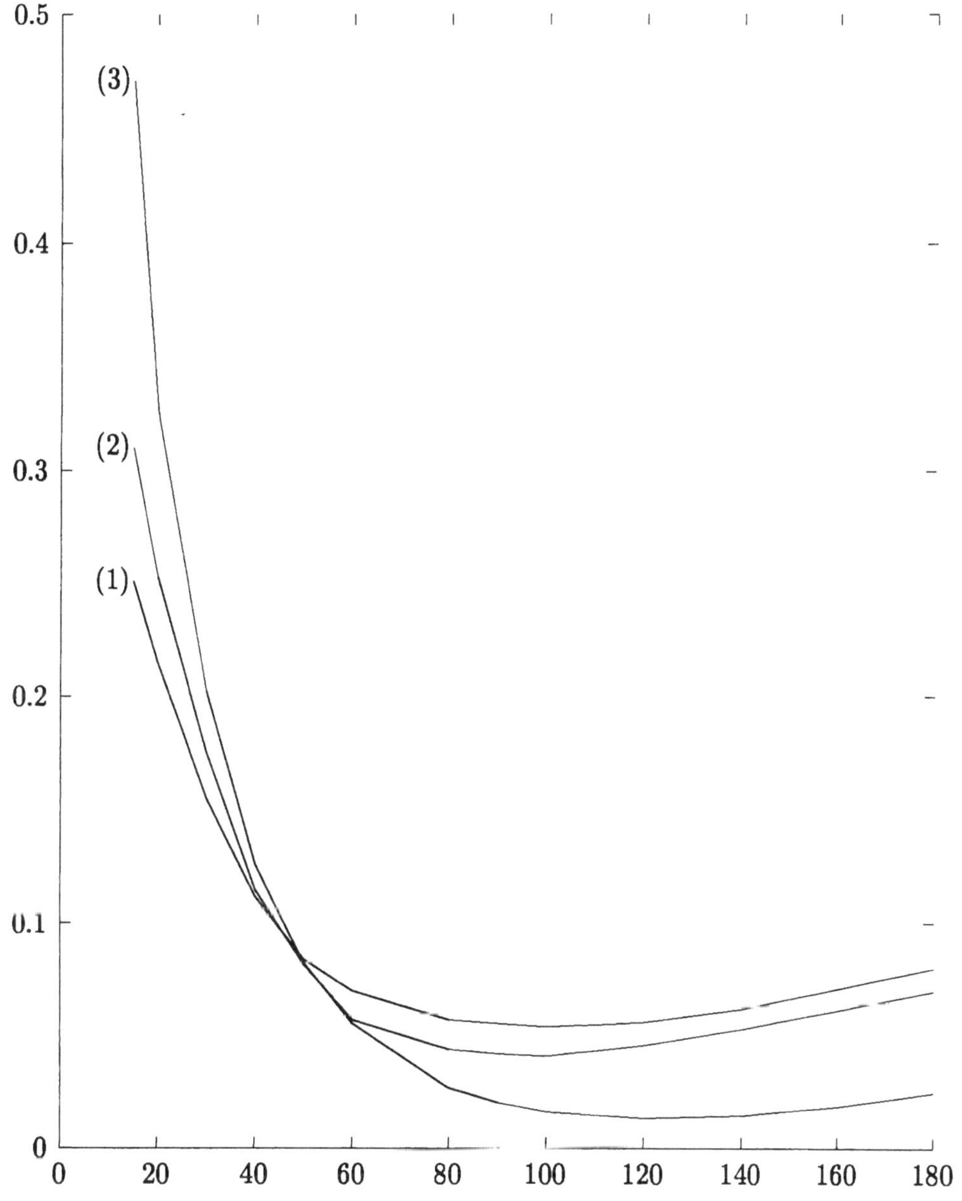

Figure 6: Normed indicatrix of brightness (1), scattering indicatrix (2), and aerosole indicatrix (3) ($\tau = 0.259$, $\lambda = 0.547\ \mu$

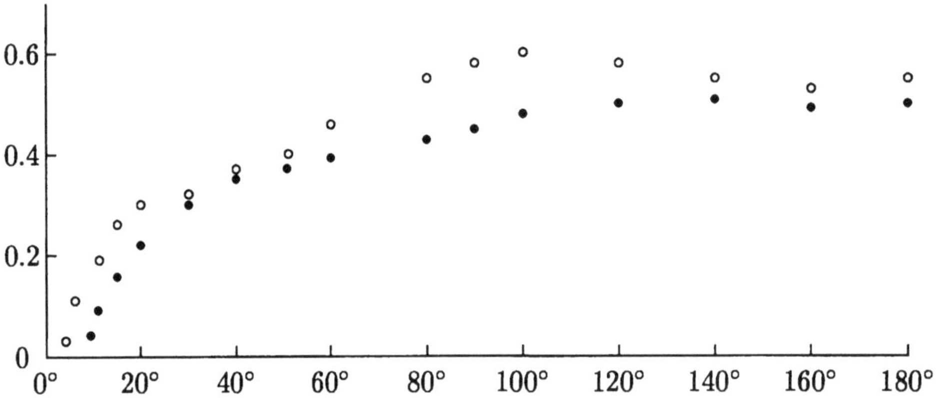

Figure 7: Angular dependence of the proportion of multiscattered component in the total intensity of the scattered radiation: ∘ — $\tau = 0.259$, $\lambda = 0.547\ \mu$; • — $\tau = 0.099$, $\lambda = 2.42\ \mu$

4.4. NUMERICAL RESULTS

The studied spectrum region was 0.405–2.42μ. In the short-wave region ($\lambda \leq 0.547\mu$) measurement data have been used with $\tau \leq 0.3$. Greater values of τ required a significant increase in measurement accuracy and computation time.

Apparently, there exist a limiting value of τ for which the proposed method is still effective (see the next section). For greater values of τ new ways for solving the problem should be developed (see Chapter 5).

As a result of the described calculations, 19 phase functions have been estimated. The averaged aerosole indicatrices for each studied wavelength are presented in Table 2 along with the averaged coefficients of asymmetry Γ_a. Figure 6 shows the forms of the normed radiation intensity indicatrices and aerosole indicatrices for $\tau = 0.259$.

One can see from this figure that the form of scattering indicatrix differs significantly from that of brightness indicatrix. Figure 7 shows a change in multiscattered radiation. The share of aerosole scattering in the total scattering is shown in Figure 8. In the case of $\tau = 0.259$ at the scattering angle $100° - 140°$ it makes up to 10%. This shows how difficult it is to separate that component.

Dependence of the form of the aerosole scattering indicatrix is of great

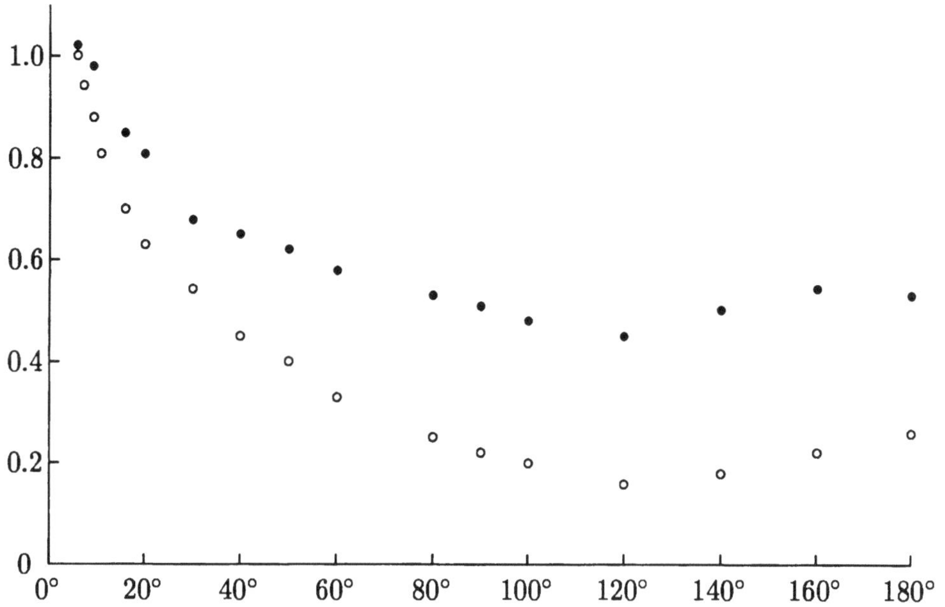

Figure 8: Angular dependence of the proportion of aerosole-scattered component in the total intensity of the scattered radiation: ∘ — $\tau = 0.259$, $\lambda = 0.547\ \mu$; • — $\tau = 0.099$, $\lambda = 2.42\ \mu$

interest. It can be observed from the data in Table 2, where the values of asymmetry are presented.

According to these data, the value of asymmetry Γ_a decreases slightly as the wavelength increases, reaches its minimum at $\lambda = 1.01\mu$, and then starts increasing. The decrease of Γ_a may be explained by a changing relation between the wavelength and the particle dimension. The increase in asymmetry may be related to the refraction parameter of large particles, which plays an important role in IR-region of the spectrum.

We note that the obtained results show the effectiveness of the proposed method. The modern problems of interpretation of measurement data from a spacecraft require more precise values of the optical parameters. They need not a local but an integral phase function related to the whole atmosphere.

Table 2: Aerosole phase functions and their coefficients of asymmetry

φ°	λ, μ				
	0.547	0.74	1.01	1.58	2.42
0	80.1000	62.6000	36.2000	18.5000	11.7000
1	40.3000	35.2000	22.0000	12.8000	9.1000
2	14.6000	16.6000	11.9000	8.6000	7.2300
3	5.6900	8.3800	7.1600	6.1300	5.6800
4	4.4900	5.4900	5.0900	4.6600	4.4900
5	2.6900	3.4400	3.5100	3.5400	3.6000
6	2.0600	2.6300	2.6500	2.8000	2.9400
8	1.2400	1.5100	1.6000	1.8800	2.0300
10	0.8590	0.9740	1.0700	1.3300	1.4900
15	0.4710	0.4720	0.5460	0.6500	0.7460
20	0.3260	0.2980	0.3350	0.3940	0.4620
30	0.2020	0.1720	0.1720	0.1740	0.1830
40	0.1260	0.1080	0.1060	0.1020	0.1020
50	0.0824	0.0709	0.0689	0.0639	0.0610
60	0.0557	0.0510	0.0489	0.0456	0.0416
70	0.0377	0.0363	0.0356	0.0325	0.0284
80	0.0269	0.0280	0.0279	0.0256	0.0224
90	0.0204	0.0219	0.0228	0.0206	0.0179
100	0.0164	0.0181	0.0196	0.0176	0.0150
110	0.0142	0.0164	0.0176	0.0162	0.0131
120	0.0134	0.0151	0.0176	0.0153	0.0120
130	0.0134	0.0148	0.0177	0.0148	0.0121
140	0.0147	0.0147	0.0194	0.0153	0.0125
150	0.0168	0.0152	0.0202	0.0165	0.0132
160	0.0186	0.0160	0.0213	0.0181	0.0146
170	0.0210	0.0179	0.0230	0.0194	0.0158
180	0.0248	0.0210	0.0254	0.0212	0.0174
Γ_a	9.0	8.8	7.2	8.5	10.7

4.5. STUDY OF CONVERGENCE

In this section we study convergence of the iterative method for identification of the phase function described in Section 2.

First we consider a notation that refers to the identification method. The integral operator for the transport equation in a plane slab has the form

$$[Kf](z,w) = \begin{cases} \int\limits_{\Omega} d\omega' \frac{g(\omega',\omega)}{2\pi} \int\limits_{0}^{z} dt \frac{\tau}{\mu} \exp\left(\frac{-\tau(z-t)}{\mu}\right) f(t,\omega'), & \mu > 0, \\ f(z,\omega), & \mu = 0, \\ \int\limits_{\Omega} d\omega' \frac{g(\omega',\omega)}{2\pi} \int\limits_{0}^{z} dt \frac{\tau}{-\mu} \exp\left(\frac{-\tau(z-t)}{-\mu}\right) f(t,\omega'), & \mu < 0. \end{cases}$$

Here $\mu = c$, integration limits with respect to t are equal to:

$$0,\ z \quad \text{if} \quad \mu < 0;$$
$$z,\ 1 \quad \text{if} \quad \mu > 0.$$

We denote by $I_\tau^i(g)$ and $F_\tau^i(g)$ the respective intensities of the single- and multiscattered radiation observed in the i-th direction. Let

$$I_\tau = (I_\tau^1, \ldots, I_\tau^n),$$
$$F_\tau = (F_\tau^1, \ldots, F_\tau^n).$$

For the function $f(z,w)$ $(0 \le z \le 1)$, we introduce the norm

$$\|f\| = \int\limits_{\Omega} d\omega \int\limits_{0}^{1} dz |f(z,w)|.$$

The recurrent formula for each iteration has the form (see the preceding sections):

$$g^{k+1} = \frac{I_\tau(g^*) + F_\tau(g^*) - F_\tau(g^k)}{S(I_\tau(g^*) + F_\tau(g^*) - F_\tau(g^k))} \equiv G_\tau(g^k).$$

Here G_τ is a nonlinear operator of iteration and S is the linear functional representing the norming constant of the vector $x = (x_1, \ldots, x_n)$, namely

$$S(x) = \sum_{i=1}^{n-1} \frac{x_i + x_{i+1}}{2} (\mu_i - \mu_{i+1}),$$

$$g^k - g^* = \delta^k, \qquad g^{k+1} - g^* = G_\tau(g^k) - G_\tau(g^*).$$

Theorem 1. *Let $\Omega_R = \{\|\delta\| \leq R\}$, $R > 0$, and let all the coordinates of the vectors $g^* + \delta$, $\delta \in \Omega$, be bounded below by a positive number d. Then for a sufficiently small τ the estimate*

$$\|G_\tau(g) - G_\tau(g^*)\| \leq q\|d\|, \quad q < 1, \tag{4.1}$$

holds uniformly on the set Ω_R.

Proof. The proof leans upon two auxiliary statements.

Lemma 1. *If $f(z,w) > 0$ and $g = g^* + \delta$, $\delta \in \Omega_R$, then*

$$\|Kf\| \leq C_\tau \|f\|, \tag{4.2}$$

where $C_\tau \to 0$ as $\tau \to 0$.

Proof. By definition we have

$$\|Kf\| = \int_\Omega d\omega \int_0^1 dz \;\; [Kf](z,w).$$

Let us estimate the integral with respect to z, provided that $\mu > 0$. We have

$$\begin{aligned} J &= \int_0^1 dz \;\; [Kf](z,w) \\ &= \int_0^1 dz \int_\Omega d\omega' \frac{g(\omega',\omega)}{2\pi} \int_0^z dt \, \frac{\tau}{\mu} \exp\left(-\frac{\tau(z-t)}{\mu}\right) f(t,\omega'). \end{aligned}$$

Changing the order of integration, we obtain successively

$$\begin{aligned} J &= \int_\Omega d\omega' \frac{g(\omega',\omega)}{2\pi} \int_0^1 dz \int_0^z dt \, \frac{\tau}{\mu} \exp\left(-\frac{\tau(z-t)}{\mu}\right) f(t,\omega') \\ &= \int_\Omega d\omega' \frac{g(\omega',\omega)}{2\pi} \left[\int_0^1 dt \int_0^t dz \, \frac{\tau}{\mu} \exp\left(-\frac{\tau(z-t)}{\mu}\right) f(t,\omega') \right]. \end{aligned}$$

Let us estimate the integral in brackets:

$$\begin{aligned}
&\int_0^1 dt \int_t^1 dz \frac{\tau}{\mu} \exp\left(-\frac{\tau(z-t)}{\mu}\right) f(t,\omega') \\
&= \int_0^1 dt\, f(t,\omega') \int_t^1 dz \frac{\tau}{\mu} \exp\left(-\frac{\tau(z-t)}{\mu}\right) \\
&= \int_0^1 dt\, f(t,\omega') \left(1 - \exp\left(-\frac{\tau(1-t)}{\mu}\right)\right) \\
&< \int_0^1 dt\, f(t,\omega') \left(1 - \exp\left(-\frac{\tau}{\mu}\right)\right).
\end{aligned}$$

Thus, for $\mu > 0$ we have the inequality

$$J < \int_\Omega d\omega' \frac{g(\omega',\omega)}{2\pi} \left[\int_0^1 dt \;\; f(t,\omega')\right] \left(1 - \exp\left(-\frac{\tau}{\mu}\right)\right).$$

A similar estimate can be easily obtained for J at $\mu < 0$, with $|\mu|$ standing instead of μ. Hence, we obtain the estimate

$$\begin{aligned}
\|Kf\| &= \int_\Omega J d\omega \\
&< \int_\Omega d\omega \int_\Omega d\omega' \frac{g(\omega',\omega)}{2\pi} \left(1 - \exp\left(-\frac{\tau}{\mu}\right)\right) \left[\int_0^1 dt \;\; f(t,\omega')\right] \\
&= \int_{|\mu|\le\varepsilon} + \int_{|\mu|>\varepsilon}.
\end{aligned}$$

Let us estimate the first integral. We use the inequality

$$1 - \exp\left(-\frac{\tau}{\mu}\right) < 1.$$

Hence,

$$\int\limits_{|\mu|\le\varepsilon} < \int\limits_{|\mu|\le\varepsilon} d\omega \int\limits_{\Omega} d\omega' \frac{g(\omega',\omega)}{2\pi} \left[\int\limits_0^1 dt\, f(t,\omega')\right]$$

$$= \int\limits_{\Omega} d\omega' \left[\int\limits_0^1 dt \quad f(t,\omega')\right] \left[\int\limits_{|\mu|\le\varepsilon} d\omega \frac{g(\omega',\omega)}{2\pi}\right]$$

$$\le D_\varepsilon \left[\int\limits_{\Omega} d\omega' \int\limits_0^1 dt \quad f(t,\omega')\right] = D_\varepsilon \|f\|,$$

where

$$D_\varepsilon = \max_{g^*\in S_R} \max_{\Omega} \left[\int\limits_{|\mu|\le\varepsilon} d\omega \frac{g(\omega',\omega)}{2\pi}\right] \to 0.$$

Let us estimate the second integral. Note that it follows from $|\mu| < \varepsilon$ that

$$1 - \exp\left(-\frac{\tau}{\mu}\right) < 1 - \exp\left(-\frac{\tau}{\varepsilon}\right)$$

and it follows from

$$\int\limits_{\Omega} d\omega \frac{g(\omega',\omega)}{2\pi} = 1$$

that

$$\int\limits_{|\mu|>\varepsilon} d\omega \frac{g(\omega',\omega)}{2\pi} < 1.$$

Thus, we have

$$\int\limits_{|\mu|>\varepsilon} d\omega < (1-\exp(-\tau/\mu)) \int\limits_{|\mu|>\varepsilon} d\omega \int\limits_{\Omega} d\omega' \frac{g(\omega'\cdot\omega)}{2\pi} \left[\int\limits_0^1 dt f(t,\omega')\right]$$

$$= (1-\exp(-\tau/\mu)) \int\limits_{\Omega} d\omega' \left[\int\limits_0^1 dt f(t,\omega')\right] \int\limits_{\Omega} d\omega' \frac{g(\omega'\cdot\omega)}{2\pi}$$

$$< (1-\exp(-\tau/\mu)) \left[\int\limits_0^1 dt f(t,\omega')\right] = H_{\tau,\varepsilon}\|f\|,$$

where $H_{\tau,\varepsilon} \to 0$ as $\tau \to 0$. Finally, we find that the estimate

$$\|Kf\| \le (H_{\tau,\varepsilon} + D_\varepsilon)\|f\| \tag{4.3}$$

holds uniformly on the set $\{g^* + \Omega_R\}$. It is clear that by choosing successively the numbers ε and τ the multiplier $H_{\tau,\varepsilon} + D_\varepsilon$ can be made as small as is wished. This proves the validity of the estimate (2), where $C_\tau = H_{\tau,\varepsilon} + D_\varepsilon$. □

Note that the function f is assumed to be positive everywhere and therefore, the inequality (3) cannot be considered as an estimate for $\|K\|$. However, the inequality $\|Kf\| \le C_\tau \|f\|$ holds.

Lemma 2. *There exists the limit*

$$\lim_{\tau \to 0} \frac{D_j F_\tau(g^* + \delta)}{I(\tau)} \to 0$$

as $g^* + \delta \in \Omega_R$.

Proof. We use the expression for the statistical estimate of the functional $D_j F$ at the point of collision (see Antyufeev (1980)). Besides, we take into account that $\min g_i \ge d$ on the set $\{\Omega_R\}$ and hence

$$\max \frac{1}{g} = \frac{1}{d} < \infty.$$

We have

$$D_j F_\tau(g^* + \delta) \le C \cdot \|\psi\| \cdot \left[\frac{C_\tau}{d} + \frac{C_\tau}{d^2} + \dots\right]$$
$$= C \cdot \frac{\|\psi\|}{d} \cdot \frac{C_\tau(2 - C_\tau)}{(1 - C_\tau)^2} \to 0$$

as $\tau \to 0$. Here $C_\tau \to 0$. Thus, we proved that

$$\lim_{\tau \to 0} \frac{\operatorname{grad} F_\tau(g^* + \delta)}{I(\tau)} = 0$$

as $g^* + \delta \in \Omega_R$. □

Now, using Lemmas 1 and 2, we prove the estimate (1). According to the mean-value theorem, for $\delta \ne 0$ we have

$$F_\tau(g^* + \delta) - F_\tau(g^*) = \frac{\partial F_\tau}{\partial u}(g^* + \delta\theta)h,$$

where $h = \|\delta\|$, $u = \delta/\|\delta\|$, and $0 < \theta < 1$. Hence,

$$\frac{\operatorname{grad} F_\tau(g^* + \delta)}{hI(\tau)} = \operatorname{grad} F_\tau(g^* + \delta\theta)u/I(\tau) \to 0$$

at $\delta \in \Omega_R$, $\delta \neq 0$.

Thus, for small τ the following estimate holds on the set $\{\Omega_R\}$:

$$\frac{F_\tau(g^* + \delta) - F_\tau(g^*)}{I(\tau)} \leq \varepsilon\|\delta\|,$$

where $\varepsilon > 0$ is an arbitrary prescribed number. The functional S is prescribed on a finite-dimensional space and hence it is bounded. Let N be its norm. Then

$$\|G_\tau(g) - G_\tau(g^*)\| \leq C \cdot (\varepsilon\|\delta\| + \varepsilon N\|g^*\|\|\delta\|) = C \cdot \varepsilon\|\delta\|, \tag{4.4}$$

where $\varepsilon > 0$ can be made as small as is wished as $\tau \to 0$. Thus, the estimate (4) holds for sufficiently small τ.

Hence, if $g^k \in \Omega_R$ is the k-th iteration, then

$$\|g^{k+1} - g^*\| \leq \|g^k - g^*\| \leq q^n\|g^0 - g^*\|$$

and therefore, $g^{k+1} \in \Omega_R$. Thus, the process convergence is proved.

Finally, we make two notes:

1. For a given phase function g we can point out explicitly the interval $[0, T]$ of optical thickness τ of the layer for which the recovery process converges.

2. The proof provides only the sufficient convergence condition and therefore, the process can also converge outside the interval. The proof, however, confirms the results of the numerical experiments, which show that the rate of convergence decreases as τ grows and the phase function becomes more elongated. For sufficiently large τ the process does not seem to converge even theoretically (i. e. even in the absence of computer errors). However, as shown above, the field of algorithm application can be extended using regularization (see Chapter 5).

Chapter 5.

Regularization in solving inverse problems of atmosphere optics

Ground, atmospheric, and satellite observations yield information on the angular distribution of brightness for the scattered solar radiation. Relying on this information, we can pose and solve inverse problems on determining optical parameters of the atmosphere.

In Chapter 1 we have considered a few developments that improve the Monte Carlo algorithms for solving direct problems in atmosphere optics. Modifications of the Monte Carlo method have been proposed to reduce computation costs of the used algorithms.

In this chapter we consider two inverse problems:

1. The recovery of altitudinal variations of the scattering coefficient by observing the brightness of the diurnal terrestrial horizon. The iterative algorithms for solving the problems are briefly presented. The convergence of the algorithms is studied as well as various ways of their mathematical optimization.

2. The identification of the phase function of the angular light diffusion by observing the brightness in the solar almucantar.

Numerical experiments have shown that the quality of identification of the scattering indicatrix is deteriorating as the optical thickness τ of the atmosphere layer increases. Peaks appear on the graph of the recovered function, which is not typical of the physical function. This is due to the fact that the problem becomes ill-posed. Various means of tackling the difficulty were used. For example, the phase function can be identified at a less number of nodes and then interpolated on the intervals by using an interpolation formula. But arbitrary change of the curve form and the function smoothing without regard for its nature also result in distortions and do not ensure the quality of recovery. Another direction of the study for the improvement of the algorithms for solving inverse problems is related to the use of regularization.

It should be noticed that the usual regularization that apply smoothing functionals cannot be used here. The matter is that the desired functions are not smooth. The scattering coefficient as a function of the altitude over the earth's surface changes in a very confused way. Behavior of the phase function $g(\mu)$ in the neighborhood of the point $\mu = 1$ looks like a discontinuity of the second kind. The phase function abruptly increases as $\mu \to 1$ (we have mentioned it in the second chapter).

But the conventional regularization methods are just intended to remove such "defects" in the function behavior. Therefore, here we should use another type of regularization in order to take into account characteristic properties of the function to be estimated. In contrast to the commonly applied regularization procedure we will not seek a smoothing functional, but use an additional information on the needed function of analytical or statistical kind.

By the analytical information we mean the following. In both considered problems values of certain linear functionals of the desired function are known. We may regard them as an additional condition which helps to reduce the estimation error. The statistical information means a known correlation function of an ensemble from which the desired function should be chosen.

First we consider in brief the statement and algorithm for solving the problem. Then we describe methods of regularization.

Besides, it should be noted that the usual ways of regularizing, based on smoothing functionals, cannot be applied here. The reason is that the sought function $g(\mu)$ is not smooth in the common sense. The physical scattering phase function $g(\mu)$ increases sharply as $\mu \to 1$, which is like the second-kind break near the point $\mu = 1$.

Here we describe briefly the inverse problems to be studied. See Antyufeev and Mikhailov (1976), Antyufeev and Nazaraliev (1988a) for more details.

5.1. RECOVERY OF ALTITUDINAL VARIATIONS OF THE SCATTERING COEFFICIENT

A parallel flux of solar radiation is incident on the upper atmosphere surrounding the Earth. The atmosphere is divided into n layers throughout its height, with the scattering coefficient being constant inside each layer.

A radiation detector is located at a point S above the outer atmosphere of the Earth. It measures the intensity of scattered solar radiation coming into the point in N different directions, $N > n$ (see Figure 9). The problem is to recover the scattering coefficients $\sigma_1, \sigma_2, \dots \sigma_n$ in all the layers by using the measured intensities $I_1, I_2, \dots, I_N$. All the other optical parameters are assumed to be known.

From the mathematical point of view the problem is reduced to the solution of the system of N nonlinear equations in n unknowns

$$\begin{aligned} I_1(\sigma_1, \dots, \sigma_n) &= I_1^* \\ I_2(\sigma_1, \dots, \sigma_n) &= I_2^* \\ &\dots\dots\dots \\ I_N(\sigma_1, \dots, \sigma_n) &= I_N^* \end{aligned}$$

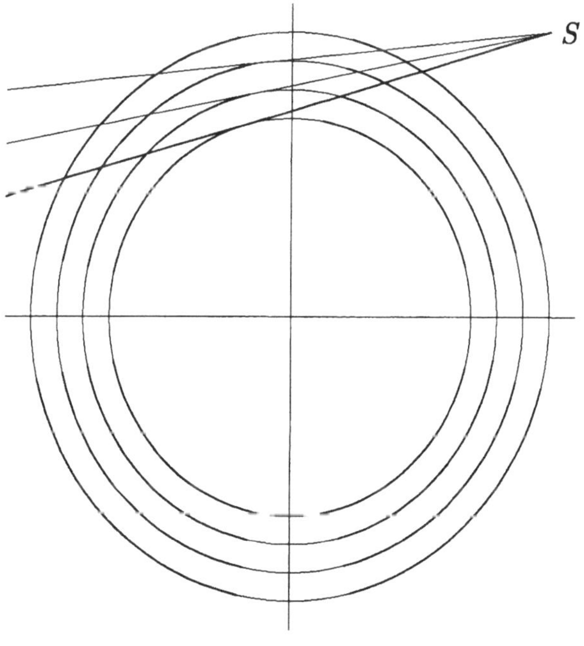

Figure 9: Scheme of observation from the Earth satellite

Here $I_1, I_2, \ldots, I_N$ are the observed intensities. The system is solved by the Newton iterative method (the method of tangent mapping), namely the system of linear equations is solved at the k-th step:

$$\begin{aligned} D_{11}\delta_1 + \ldots + D_{1n}\delta_n &= I_1^* - I_1^{(k)} \\ D_{21}\delta_1 + \ldots + D_{2n}\delta_n &= I_2^* - I_2^{(k)} \\ &\ldots\ldots\ldots \\ D_{N1}\delta_1 + \ldots + D_{Nn}\delta_n &= I_N^* - I_N^{(k)}, \end{aligned} \tag{5.1}$$

where

$$\delta_n = \sigma_m^{(k)} - \sigma_m^{(k-1)}, \qquad I_m^{(k)} = I_m(\sigma_1^{(k)}, \ldots, \sigma_N^{(k)}), \qquad D_{ij} = \frac{\partial I_i}{\partial \sigma_j},$$

and $\sigma_m^{(k)}$ is the k-th approximation to the sought coefficient σ_m^*. Since $N > n$, the system, in general, is overdetermined. It can be solved by the least square method.

Marchuk was the first to formulate the problem this way (Marchuk, 1964; Marchuk and Mikhailov, 1967).

5.2. ANALYTICAL REGULARIZATION

The optical thickness of the atmosphere is often known in the problem on recovering the altitudinal variations of the scattering coefficient. For the considered atmosphere model this value is equal to

$$\tau = \sum_{i=1}^{n} \sigma_i h_i,$$

where h_i is the geometric thickness of the i-th layer. We can use this condition to regularize the problem by calculating the conditional extremum.

Instead of solving the system of equations (5.1) with the matrix D, we will search for the vector $\delta^{(k)}$, which minimizes the functional

$$\min |D\delta^{(k)} - (I^* - I^{(k)})|^2,$$

provided

$$\sum_{i=1}^{n} \delta_i^{(k)} h_i = \tau_0 - \tau^{(k)},$$

where

$$\tau^{(k)} = \sum_{i=1}^{n} \sigma_i^{(k)} h_i.$$

This problem can be solved by the method of Lagrange multipliers. Here the Lagrange function is of the form:

$$F(\delta^{(k)}, \lambda) = |D\delta^{(k)} - (I^* - I^{(k)})|^2 - \lambda \left(\sum_{i=1}^{n} \delta_i^{(k)} h_i - (\tau_0 - \tau^{(k)}) \right).$$

To find the vector $\delta^{(k)}$, which minimizes the functional, we solve the system of equations

$$\frac{\partial F}{\partial \delta_i^{(k)}} = 0, \quad i = 1, \dots, n$$

$$\sum_{i=1}^{n} \delta_i^{(k)} h_i - (\tau_0 - \tau^{(k)}) = 0.$$

Thus, at each iteration step we solve a system of linear equations of the $(n+1)$-th order for the unknowns $\delta_1^{(k)}, \dots, \delta_n^{(k)}$, λ. The matrix of this system is of the form

$$\mathrm{A} = \left[\begin{array}{cccc|c} & & & & h_1 \\ & & & & h_2 \\ & & D^*D & & \vdots \\ & & & & h_n \\ \hline h_1 & h_2 & \cdots & h_n & 0 \end{array}\right]$$

and the right-hand side of the system is a vector with the components

$$b_1, b_2, \dots, b_n, \tau - \tau^{(k)},$$

where

$$(b_1, b_2, \dots, b_n) = b = D^*(I^* - I^{(k)}).$$

Numerical experiments have shown that the matrix A is better conditioned than the matrix D^*D.

5.3. REGULARIZATION IN ESTIMATING THE PHASE FUNCTION

The same approach may be used in identifying the scattering phase function.

First, recall the iterative algorithm described in detail in Chapter 4.

Here we model the atmosphere as a plane homogeneous layer. A detector is located on the earth's surface, it measures intensity of the scattered solar radiation. Zenith angles of the observation directions are equal to each other and to the solar zenith angle $\theta_{\odot}$:

$$\theta = \theta_{\odot}.$$

Azimuth angles between vertical planes of observation directions and a fixed plane of the solar vertical vary in the interval $0° < \varphi < 180°$.

The single-scattering angle α in that scheme varies in the limits $0 < \alpha < 2\theta < 180°$. Here the angle α is equal to angle between the observation direction and the solar direction.

Values of the other parameters of the atmosphere model — coefficients of molecular and aerosol scattering, albedo — are assumed to be known.

The problem is posed as follows. Estimate the phase function knowing the measured values $I_1^*, \ldots, I_m^*$. The phase function can be defined by its values $g_1, g_2, \ldots g_n$ at the node points. Thus, from the mathematical point of view the problem may be reduced to the problem of determining the values $g_1, g_2, \ldots g_n$. Take a note that because of the choice of the observation direction we have $2\theta < 180°$. Therefore, the number m of the observation direction (i. e. the number of given values) is less then the number n of unknown values.

The method for estimating the phase function is iterative. It is based on separation of single-scattered radiation from the observed radiation flux. Let $I_1, \ldots, I_m$ be the values of radiation intensity observed at the detector point:

$$I_k = I_k(g_1, \ldots, g_n).$$

Then the formula for the corresponding single-scattered radiation intensity is

$$J_k = C \cdot g_k,$$

where

$$C = \frac{\exp(-\tau)}{2\pi} \cos\theta_{\odot},$$

τ is the optical thickness of the atmosphere in the observation direction. τ and C are independent of direction number since all the zenith angles are equal to each other. We obtain an iterative process from the approximate relation

$$I_k^* - I_k \approx J_k^* - J_k = C(g_k^* - g_k).$$

In the r-th step we solve a trivial diagonal underdetermined system of linear equations

$$C\delta_k = I_k^* - I_k^{(k)}, \quad m = 1, \dots, n, \tag{5.2}$$

where

$$\delta_m = g_m^{(k)} - g_m^{(k-1)}, \qquad I_m^{(k)} = I_m(g_1^{(k)}, \dots, g_n^{(k)}),$$

$g_m^{(k)}$ is the k-th approximation to the m-th component of the sought vector $g^* = (g_1^*, \dots, g_n^*)$. Since $r < n$, at each iteration we recalculate only r components of n. We call the rest $(n-r)$ components a "tail" of the phase function. To find it, we take an "averaged" phase function, multiply its r-th to n-th components by such a factor that its r-th component equals to the r-th component of the estimated phase function. Then we put the unknown components $g_r, \dots, g_n$ equal to the relevant components of the "averaged" phase function.

Of course, this estimation looks rather strange. Below we use a regularization to get rid of this defect of the algorithm of estimation.

For the regularization procedure we use the norming relation

$$\sum_{i=1}^{n-1} \frac{g_i + g_{i+1}}{2}(\mu_i - \mu_{i-1}) = 1,$$

which can be rewritten as

$$\sum_{i=1}^{n} k_i g_i = 1, \tag{5.3}$$

where

$$k_1 = \frac{\mu_1 - \mu_2}{2},$$
$$k_i = \frac{\mu_{i-1} - \mu_{i+1}}{2}, \quad i = 2, \dots, N-1,$$
$$k_N = \frac{\mu_{N-1} - \mu_N}{2}.$$

In fact, each iteration is reduced to solving a diagonal system of linear equations. So we can use the approach we have used to estimate the scattering coefficient. We will use the Lagrange method to find a vector g which minimizes the functional

$$F(\delta^{(k)}, \lambda) = |C\delta^{(k)} - (I^* - I^{(k)})|^2 - \lambda\left(\sum_{i=1}^{n} \delta_i^{(k)} h_i - (\tau_0 - \tau^{(k)})\right).$$

The calculating procedure may be reduced to solving a system of linear equations with the matrix

$$\mathrm{A} = \left[\begin{array}{c|c} C^*C & \begin{matrix} k_1 \\ k_2 \\ \cdot \\ \cdot \\ \cdot \\ k_n \end{matrix} \\ \hline \begin{matrix} k_1 & k_2 & \cdot & \cdot & \cdot & k_n \end{matrix} & 0 \end{array}\right]$$

Here the coefficients k_i are taken from the norming condition.

5.4. STATISTICAL REGULARIZATION

We have statistical information on the sought solution and we can use it in order to regularize the problem. Further we consider the method of statistical regularization proposed by Turchin (Kozlov, Malkevich, and Turchin, 1970; Turchin, 1967; Turchin and Nozik, 1969). This method extends the area of solvability of the posed problem. The method can be used for solving systems of linear equations with *a priori* known distribution of the sought vector (we have already mentioned that at each step the considered iterative algorithms involve the solution of a linear system). The sought vector is generally considered to have the normal distribution with a known correlation matrix. Such matrices C are known for the altitudinal variations of the scattering coefficient (Kondrat'yev, Marchuk, *et. al.*, 1977). There are no such matrices for the scattering phase function since different investigators use different descriptions of the phase function because they use different nodes for approximations. However, at our disposal we have a big ensemble of indicatrices and using it we can find such a matrix. Having taken the same approximation nodes, we find the correlation matrix C for the stochastic vector $g = (g_1, \dots, g_n)$.

The correlation matrix describes the *a priori* distribution of the sought vector and the solved system of linear equations determines its "model" distribution. Note that the right-hand side of the system is considered here as a stochastic vector (rather than a deterministic one) with independent and normally distributed components for which the measurement errors are known. Now, using the Bayes theorem, we can find the *a posteriori* distribution of the sought vector. All the calculations are carried out explicitly and analytically, since all the distributions are assumed to be normal. The expectation of the *a posteriori* distribution is taken as a regularized solution.

The algorithm of solution can be represented as follows. If we know the correlation matrix C, then instead of the initial system of linear equations

$$D\delta = b$$

with the matrix D we are to solve the linear system

$$(D^*WD + C^{-1})\delta = D^*Wb$$

with the regularized matrix $D^*WD + C^{-1}$, where W is a diagonal matrix with diagonal elements

$$w_{ii} = \frac{1}{\varepsilon_{ii}^2},$$

and ε_{ii} is the mean-statistical error of the component b_i of the vector b.

5.5. APPLICATION OF STATISTICALLY ORTHOGONAL EXPANSIONS

Here we consider the piecewise linear approximation of the scattering phase function. A piecewise linear approximation can be replaced by the expansion of the function in terms of the basis of "peak" piecewise linear functions, with the number of these functions being equal to the number of nodes and the coefficients of expansion being equal to the values of the function at the approximation nodes.

Thus, considering the piecewise linear approximation to the phase function with n nodes, we, in fact, expand the phase function into a series consisting of n terms. Hence, when solving the inverse problem, we should recover n unknown parameters, namely the components of the sought vector $g = (g_1, \dots, g_n)$. Regularizing the problem with the use of the method of selecting the principal components (Hotelling, 1933; Anderson, 1963), we can reduce this number n.

The reason is that we have a statistical ensemble of indicatrices. Therefore, the recovered vector can be considered as a stochastic vector belonging to the given statistical ensemble and having the known covariation matrix C. Each component g_i has a definite variance in the ensemble. If we considered the representation of the vector g in another system of coordinates, then, generally speaking, the variances of the projections of the vector g onto the new axes would differ. The method of principal components just consists in selecting successively such mutually orthogonal directions for which the variances are the largest. These directions turn out to be eigenvectors for the covariation matrix C of the stochastic vector g and can be easily found explicitly (Hotelling, 1933). The corresponding basis is called the basis of statistically orthogonal functions. For the successively remaining directions the variances decrease.

Represent the vector g as the sum

$$g = \overline{g} + \widetilde{g},$$

where $\overline{g}$ is a determinate vector equal to the ensemble average of g over the statistical ensemble (we know this vector). Now we search for the vector $\widetilde{g}$. Let $h_1, \dots, h_n$ be vectors of a statistically orthogonal basis. Consider the decomposition of the vector g in this basis and study its last components (with respect to the value of variance). Assume that k components are negligibly small, $k < n$, and put them equal to zero. Since the basis is orthogonal, this assumption can be written analytically as the system of n equalities

$$(\widetilde{g}, h_i) = 0, \quad i = n - k + 1, \dots, n. \tag{5.4}$$

Now let us see how these results can be applied in practice to solve the inverse problem. In the numerical experiments we have used the ensemble of scattering indicatrices based on the data, which have been put at our disposal by the Institute of Applied Geophysics (Moscow) (GOST 25645.153-90, 1991; Radiation Commission IAMAP, WMO.WCP-112, 1986). The number m of nodes in the piecewise linear approximation was equal to 23. The numerical experiments have shown that in expanding in terms of a statistically orthogonal basis we could restrict ourselves to the first 15 terms of the series. In terms of our notation we have $n = 23$ and $k = 8$. In doing so, the approximation error of the phase function at the nodes was no higher than 2%, which is less than the threshold of accuracy of the phase function identification. In fact, this implies that only 15 parameters would suffice to specify the phase function. The relations (4), together with the norming conditions (3) for the vector g, result in the system of nine linear equations

$$\sum_{j=1}^{23}(\overline{g}_j - \widetilde{g}_j)h_j^i = 0, \quad i = 16, \ldots, 23$$

$$\sum_{j=1}^{23} k_j g_j = 1.$$

These equations should be considered as additional to equations (2), which we solve at each iteration step. Since the total number of equations is more than the number of unknowns (i. e. the system is overdetermined), we can drop the artificial condition according to which "a tail" is added to the phase function (see Chapter 4).

Chapter 6.

Direct and inverse problems of radiation transfer in a plant canopy

The problem of photon transport in a plant arises in the context of optical remote sensing of vegetated land surfaces, land surface climatology and plant physiology. For instance, in the application of remote sensing from satellite-based sensors to vegetated land surfaces, an understanding of the spectral response resulting from the aggregation of leaves in a canopy and the intervening atmosphere is required. The physics of this problem is most conveniently posed in the form of a photon transport equation, the solution of which is the remote spectral measurement. This sets the context for the problems considered in this chapter. Here we present a method for calculating radiation transfer in a plant canopy and identifying canopy parameters using the data from remote measurements.

The plan of the chapter is as follows. In Section 1 we present the plant canopy model. The transport equation for radiation transfer in a model plant canopy is derived in Section 2. Section 3 gives a detailed description of modelling Markov chains for solving the transport equation. The so-called "hot spot" effect in radiation transfer in a plant canopy is discussed in Section 4. The last sections are devoted to solving some inverse problems. In Section 5 we pose some inverse problems and propose a method for solving them. In the next 6-th section we give formulas for calculating derivatives which we need to solve the inverse problems. In Section 7 a new method is proposed for solving one specific inverse problem. Finally, in Sections 8 and 9 numerical results and a synopsis are given.

6.1. PLANT CANOPY MODEL

In recent years due to a fascinating progress in the space remote sensing there is a need for the development of different models of interaction of solar radiation with a plant canopy.

We may highlight two different approaches in describing the light transfer in a plant canopy:

1) Geometrical models.
   Plant canopies are modelled as geometrical bodies with given dimensions and optical properties that are special rules for transmission and reflection. Such models are more suitable for describing inhomogeneous rarefied plant canopies.

2) Models for turbid medium.
   This is a medium filled with infinitely small "leaves" with given distribution of the normal direction and given optical properties. These models serve to describe dense inhomogeneous media with phytoelements that are very small compared with the height of the plant canopy.

In more details, there are treated geometrical models, turbid medium models, compound models and statistical models (the overview of the current models can be found in Goel (1988) and Myneny, Ross, and Asrar (1989).

The model we treat below is at the interface between the turbid medium models and the statistical models. This model allows to get rid of the disadvantages of the first and second approaches. Particularly, inclusion of vertical and horizontal inhomogeneities and dimensions of the phytoelements into the turbid medium model results in significant difficulties in calculation.

Besides, using the models of the type described in Ross and Marshak (1988) for estimating geometrical parameters of a particular plant we cannot calculate derivatives with respect to the model parameters. The suggested model allows to calculate the derivatives of the radiance brightness with respect to the optical medium parameters. We need them for solving the inverse problems of identifying the optical parameters.

So, we propose a description of the radiation regime of a plant canopy using the transport equation with due regard to the most important radiation parameters. Then we consider its solving by the Monte Carlo method. The developed approach allows to generalize the technique of solving direct and inverse problems of the atmosphere optics to the case of the plant canopies.

For simplicity we limit our study to consideration of a homogenous plant canopy. In the next section we describe the integro-differential transport equation and the relevant integral equation of the second kind. Then, in the following sections we consider its solving by the Monte Carlo method. After that we include leaves dimension into the plant canopy model to take the "hot spot" effect into account.

The radiation transfer process in a plant canopy differs from the "conventional" one by the direction probability density, which cannot be defined as a function of one variable because of the lack of symmetry over the azimuth angle of the medium (see the formula for $P(\omega' \to \omega)$).

Furthermore, in a plant canopy the extinction coefficient depends not only on points of the geometrical space, but on the photon flight direction as well (see the formulas for $G(\omega)$ and $\sigma(t, \omega', \omega)$).

Next, although dimensions of the phytoelements are negligibly small, they have certain finite dimensions in a real medium. This causes the so-called "hot spot" effect, which may be taken into account by using a special formula for the coefficient $\sigma(t, \omega', \omega)$.

We will treat a slab homogeneous plate medium with a leaf area index H (which is similar to optical thickness). Its upper surface is illuminated by parallel rays of solar radiation. Its lower (bottom) surface is Lambertian, that is, it reflects the incident radiation according to the Lambertian angle distribution.

Below we give a mathematical description of the process of radiation transfer in such a plant canopy layer using the transport equation. The integro-differential equation is reduced to an integral equation of the second kind. For simplicity, we limit our consideration to the case of a homogeneous medium. However, all the conclusions could be easily referred to the case of inhomogeneous medium. Besides, here we do not take into account the diffusion component of the solar radiation scattered in the atmosphere.

6.2. TRANSPORT EQUATION FOR A PLANT CANOPY

Below we use the following notation:

ω: the photon direction right before the collision;

$\omega_{\odot}$: the direction of the solar radiation;

ω^*: the observation direction;

ω_L: the random vector of the normal to the leaf surface.

We will write for brevity $\omega \sim (\mu, \varphi)$, where $\mu = \cos\theta$, and θ, φ are the zenith and azimuthal angles of the direction ω, respectively.

$g(\omega_L)$: the leaf normal distribution function. Due to the symmetry of the system, the azimuthal angle is isotropic, that is, its density function is

$$g_\varphi(\varphi_L) \equiv \frac{1}{2\pi},$$

and hence

$$g(\omega_L) = \frac{1}{2\pi} g_\theta(\theta_L).$$

The density of the zenith angle is defined in the form of the two-variable function

$$g_\theta(\theta_L) = \frac{2}{\pi} + b\cos 2\theta_L + c\cos 4\theta_L. \tag{6.1}$$

The mean projection of the leaf normal on a direction ω is

$$G(\omega) = \frac{1}{2\pi} \int\limits_{\Omega+} d\omega_L \; g(\omega_L)|(\omega, \omega_L)|. \tag{6.2}$$

The indicatrix of transmission-reflection of Lambertian surface of the leaf having duffuse coefficients of reflectance (r_L) and transmittance (t_L) has the form

$$f(\omega' \to \omega;\; \omega_L) = \begin{cases} r_L \dfrac{|(\omega', \omega_L)|}{\pi}, & (\omega, \omega_L)\cdot(\omega', \omega_L) > 0, \\ t_L \dfrac{|(\omega', \omega_L)|}{\pi}, & (\omega, \omega_L)\cdot(\omega', \omega_L) < 0 \end{cases} \tag{6.3}$$

The non-normed indicatrix of a volume element is defined as

$$\Gamma(\omega' \to \omega) = \frac{1}{2\pi} \int\limits_{\Omega+} d\omega_L \, g(\omega_L)|(\omega, \omega_L)| f(\omega' \to \omega;\, \omega_L). \tag{6.4}$$

Conditional scattering density for the transition from a direction ω' to ω is defined by

$$P(\omega' \to \omega) = \frac{\Gamma(\omega' \to \omega)}{G(\omega')}. \tag{6.5}$$

It is normed by the value of the leaf albedo ω_L:

$$\omega_L \equiv \int\limits_{\Omega} d\omega \; f(\omega' \to \omega;\, \omega_L) = \int\limits_{\Omega} d\omega P(\omega' \to \omega),$$

where $\sigma_\kappa(t, \omega, \omega')$ is the extinction coefficient. Its formula takes into account the "hot spot" effect and the relative dimensions of the plates;

$t = t(z)$: the total leaf area index at the altitude z.
Optical distance from the point z to the upper boundary of the slab in a direction ω is equal to

$$r(x) = t(z)G(\omega).$$

$x = (t, \omega)$: a point of the phase space in the transport equation.

$I_\odot$: the intensity of solar radiation.

Now, let us consider the transport equation. The process of radiation transfer in a plane homogeneous slab of thickness H with a boundary bottom Lambertian surface having albedo r_s is described by a boundary value problem for the integro-differential equation:

$$-\mu\frac{\partial I}{\partial\tau} + G(\omega)I(\tau, \omega) = \int\limits_{\Omega+} d\omega\,\Gamma(\omega' \to \omega)I(\tau, \omega'), \tag{6.6}$$

$$I(0, \omega) = I_\odot\delta_{\omega_\odot}(\omega), \quad \omega \in \Omega_-,$$

$$I(H, \omega) = \frac{r_s}{\pi}\int\limits_{\Omega} d\omega'\ \mu I(H, \omega), \quad \omega \in \Omega_+.$$

Denote

$$J(\tau, \omega) = I(\tau, \omega) \cdot G(\omega).$$

The integro-differential equation (6) for the function $I(\tau, \omega)$ may be reduced in a usual way to the integral equation for the function $J(\tau, \omega)$. Here we drop the intermediate calculations and write down the final integral equation:

$$J(\tau, \omega) = \int\limits_{\Omega}\int\limits_{0}^{H} k((\tau', \omega') \to (\tau, \omega))J(\tau', \omega')d\tau' d\omega' + F(\tau, \omega), \tag{6.7}$$

where

$$k(x' \to x) = \begin{cases} k_1(x' \to x), & \mu < 0, \\ k_2(x' \to x), & \mu > 0. \end{cases} \tag{6.8}$$

Here

$$k_1(x' \to x) = \begin{cases} P(\omega' \to \omega)\dfrac{G(\omega)}{|\mu|}\exp\left(-\dfrac{G(\omega)}{|\mu|}(\tau - \tau')\right), & 0 \le \tau' \le \tau, \\ 0, & \text{otherwise}, \end{cases}$$

$$k_2(x' \to x) = \begin{cases} 0, & 0 \le \tau' \le \tau, \\ P(\omega' \to \omega)\dfrac{G(\omega)}{|\mu|}\exp\left(-\dfrac{G(\omega)}{|\mu|}(\tau - \tau')\right), & \text{otherwise}, \end{cases}$$

$$F(\tau, \omega) = J(0, \omega)\exp\left(\frac{G(\omega)}{|\mu|}\tau\right), \quad \mu < 0.$$

Note that the integral equation (7) may be derived right from physical considerations.

Rewrite the integral equation (7) in the operator form

$$J = KJ + F,$$

where the integral operator K has the kernel (8).

Here is a probabilistic interpretation of its form.

The kernel $k(x' \to x)$ gives the probability of a collision at the point x provided the preceding collision has taken place at the point x'.

Here $x = (\tau, \omega)$, where τ is the altitude of the collision point and ω is the photon travel direction right before the collision; $J(\tau, \omega)$ is the collision density in the phase space; $Q(\tau, \omega) = \pi \cdot F(\tau, \omega)/I_\odot|\mu_\odot|$ is the density of the first collisions of the photons, which leave the source in the direction $\omega_\odot$; $I(0, \omega)$ is the angle density of the source radiation.

It is known that if the spectral radius $\rho(K) < 1$, then the equation (7) has a unique solution that can be represented as a Neumann series

$$J = F + KF + K^2F + \dots . \tag{6.9}$$

We are going to calculate a functional of the solution J. Namely, we are interested in finding bidirectional reflection function $R(\omega^*)$ in the vision direction that may be expressed via the intensity $J(0, \omega^*)$ of the radiation reflected from the upper surface in the direction ω^*:

$$R(\omega^*) = \frac{J(0, \omega^*)}{G(\omega^*) \cdot I_\odot|\mu_\odot|}. \tag{6.10}$$

The expression for the quantity $J(0,\omega^*)$ may be written in the following form:

$$J(0,\omega^*) = (J, \delta_{x^*}) = \int_X J(x)\delta_{x^*}dx,$$

where δ_{x^*} is the Dirac delta function concentrated at the point $x^* = (0,\omega^*)$.

Taking into account (9), the latter expression for the quantity $J(0,\omega^*)$ may be reduced to the form

$$\begin{aligned} J(0,\omega^*) &= (F,\delta_{x^*}) + (KF,\delta_{x^*}) + (K^2F,\delta_{x^*}) + \dots \\ &= (F, K^*\delta_{x^*}) + (KF, K^*\delta_{x^*}) + (K^2F, K^*\delta_{x^*}) + \dots . \end{aligned}$$

We have taken into account that $(F,\delta_{x^*}) = 0$, since solar radiation and the reflected one are differently directed. K^* is the conjugate of the operator K:

$$(K^*\varphi)(x) = \int_X k(x' \to x)\varphi(x')dx'.$$

Hence,

$$(K^*\delta_{x^*})(\tau,\omega) = P(\omega \to \omega^*)\frac{G(\omega^*)}{\mu^*}\exp\left(-\tau\frac{G(\omega^*)}{\mu^*}\right).$$

Let us denote the right-hand side of the latter equation by $\psi(\tau,\omega)\cdot G(\omega^*)$ and the initial collision density by $Q(\tau,\omega)$:

$$Q(\tau,\omega) = \pi\cdot\frac{F(\tau,\omega)}{I_\odot|\mu_\odot|}.$$

Then

$$R(\omega^*) = (Q,\psi) + (KQ,\psi) + (K^2Q,\psi) + \dots,$$

where the contribution function

$$\psi(\tau,\omega) = \frac{1}{\mu^*}P(\omega\to\omega^*)\exp\left(-\tau\frac{G(\omega^*)}{\mu^*}\right).$$

Here ω^* is the viewing direction, the point of view being on the upper boundary $\tau = H$ of the medium, and $P(\omega\to\omega^*)$ is calculated according to the formulas (2)–(5).

For the reflection from the soil we have

$$\psi(\tau,\omega) = \frac{|\mu| r_s}{\pi G(\omega)}\exp\left(-H\frac{G(\omega^*)}{\mu^*}\right).$$

The i-th term in the sum (9) is the contribution of the photons of the i-th collisions to the estimate of the spectral radiance. The expansion (9) is consistent with the Monte Carlo algorithm, namely we model the photon trajectories $x_0^n \to x_1^n \to \ldots \to x_m^n$, where x_i^n is a point in the phase space corresponding to the n-th photon collision, and m is the number of the last collision before the escape or the absorption. Modelling is performed according to two distribution densities (the free path distribution density and the scattering direction distribution density $P(\omega' \to \omega)$), which enter the definition of the kernel $k(x' \to x)$. At each successive collision the contribution $W_i^n \cdot \psi(x_i^n)$ is added to the statistical estimate R:

$$R(0, \omega^*) \approx \frac{1}{N} \sum_{n=1}^{N} \sum_{i=1}^{m} W_i^n \cdot \psi(x_i^n),$$

where N is the number of trajectories and W_i^n is the "weight" of the n-th photon after the i-th collision.

6.3. MARKOV CHAIN MODELLING

To solve the transfer equation numerically one has to model a photon trajectory. We perform the simulation process according to the kernel k that consists of two probability densities: the probability density of the photon free path in the direction ω and the probability density of scattering in the same direction.

1. Free path length. The free path length is modelled as usual. The optical free path is

$$\tau_l = -\ln \alpha,$$

where α is a random number distributed uniformly over the interval $(0, 1)$. Hence, the optical free path depth is

$$\tau = -\frac{\mu}{G(\omega)} \cdot \ln \alpha.$$

2. Scattering direction modelling. Let ω' be the photon path direction prior to a collision. Let us model the direction ω after the collision. The density of the transition from the state ω' to ω has the form

$$P(\omega' \to \omega) = \frac{\Gamma(\omega' \to \omega)}{G(\omega')}.$$

Note that the modelling density should meet two requirements: first, it should be convenient for modelling and second, it should be universal. Being universal means that the density used in the calculation algorithm should not depend significantly on changes in real distribution. Otherwise, one has to look for a specific simulation formula for each calculation variant, and hence, the calculation algorithm is not universal. The Monte Carlo method proposes to simulate the process according to a unique convenient density and to multiply the particle "weight" at each successive collision by a corresponding factor to compensate for the statistical estimate bias. The "weight" enters the contribution function as a factor. Thus, let us represent the function $P(\omega' \to \omega)$ as a superposition of two densities:

$$P(\omega' \to \omega) = \int_{\Omega} d\omega_L P_1(\omega' \to \omega_L) P_2(\omega_L \to \omega).$$

Here P_1 and P_2 are conditional densities:

$$P_1(\omega' \to \omega_L) = \frac{1}{2\pi} g(\omega_L) \frac{|(\omega', \omega_L)|}{G(\omega')},$$

$$P_2(\omega_L \to \omega) \equiv f(\omega' \to \omega;\ \omega_L).$$

Now the direction ω can be modelled as follows: according to the density P_1 we model the leaf normal ω_L, and then, knowing ω_L, we model the direction ω according to the density $f(\omega' \to \omega;\ \omega_L)$. However, the density P_1 does not meet the two above-mentioned requirements. There are no simple mathematical formulas and the leaf normals distribution function $g(\omega_L)$ is multiparameter in the general case.

Assuming the leaf normals azimuth φ to be uniformly distributed and the leaf normals distribution with respect to φ to be chosen from the two-parameter family (1), we represent the density μ in the form

$$\frac{1}{2\pi} g(\omega_L) = \left[\frac{2}{\pi} \frac{1}{\sqrt{1-\mu^2}}\right] \left[\frac{1}{2\pi} g^*(\mu)\right], \tag{6.11}$$

where

$$g^*(\mu) = 1 + b(2\mu^2 - 1) + c(8\mu^4 - 8\mu^2 + 1).$$

Let us consider the first factor in (11) as a new density for the direction $\omega \sim (\mu, \varphi)$, where $f_\varphi(t) = 1/2\pi$ is the density with respect to φ and

$$f_\mu = \frac{2}{\pi\sqrt{1-\mu^2}}$$

is the density with respect to μ. Thus the density with respect to ω_L is universal (it is independent of the parameters b and c and convenient due to the simple modelling formulas:

$$\varphi_L = 2\pi\alpha, \qquad \mu_L = \sin\left(\frac{\pi\beta}{2}\right),$$

where α and β are independent random quantities distributed uniformly over the interval $(0, 1)$. Thus, the density $P_1(\omega' \to \omega)$ is represented as the product

$$P_1(\omega' \to \omega) = \left[\frac{2}{\pi\sqrt{1-\mu^2}}\right] \cdot \left[\frac{|(\omega', \omega_L)|}{G(\omega')} g^*(\mu)\right].$$

The first factor is a new density for ω_L, and the second one is multiplied by the particle "weight" at each successive collision. It should be noted that there are other methods of factorization, but the method we have chosen is convenient since here the unbounded factor $1/\sqrt{1-\mu^2}$ is included in the density and correspondingly excluded from the "weight", so the estimate remains bounded. An unbounded estimate would result in large calculation errors.

Knowing ω_L, we model the direction ω according to the density of photon interactions with a leaf surface. For the bi-Lambertian scattering law (3) modelling is realized according to the formulas

$$\varphi = 2\pi\alpha, \qquad \mu = \delta\sqrt{\beta},$$

where μ and φ are the coordinates of the vector ω_L in the spherical coordinate system with the ω axis, and the quantity δ is defined as

$$\delta = \begin{cases} +1, & \begin{array}{l} [(\omega', \omega_L) > 0] \text{ and } \left[\gamma > \dfrac{r_L}{r_L + t_L}\right] \\ \text{or} \\ [(\omega', \omega_L) < 0] \text{ and } \left[\gamma < \dfrac{r_L}{r_L + t_L}\right], \end{array} \\ -1, & \text{otherwise.} \end{cases}$$

Here α, β, and γ are random quantities distributed uniformly on the interval $(0, 1)$.

3. The contribution function evaluation. It is difficult to calculate the exact value of the quantity $\psi(\tau, \omega)$ after each collision, therefore, we should take its estimate randomized with respect to ω_L as the contribution function

$$\psi(\tau,\omega;\ \omega_L) = f(\omega \to \omega^*;\, \omega_L)\cdot\frac{1}{\mu^*}\exp(-\tau^*).$$

Here ω_L is the leaf normal modelled after a collision and τ^* is the optical depth of the layer from a collision point to a detector in the viewing direction ω^*.

6.4. TAKING INTO ACCOUNT THE HOT SPOT EFFECT

The model constructed above provides a good approximation of the plant canopy in which the phytoelements have negligible dimensions compared to plant heights and are given only by their normals distribution density. However, phytoelements have finite dimensions in the real vegetative cover. This fact leads to the so-called hot spot effect: a considerable increase in the reflected radiation is observed in the direction to the sun. The increase depends on the geometrical structure of both a single plant and the whole crop. The attenuation coefficient of the radiation flux in the direction ω after a collision proves to depend not only on the normals distribution $g(\omega_L)$ (see Section 1), but on the incident flux direction ω' as well. If one knows the cross-correlation coefficient of the indicator functions (that indicate the presence of plates in the directions ω and ω' in a differential layer) (Nilson and Kuusk, 1989), one can evaluate the new attenuation coefficient, which depends on phytoelement dimensions. For a homogeneous vegetative cover we have

$$\sigma_\kappa(t,\omega,\omega') = \begin{cases} G(\omega), & \mu\mu' < 0,\\ G(\omega)h_\kappa(t,\omega,\omega'), & \text{otherwise,}\end{cases}$$

where

$$h_\kappa(t,\omega,\omega') = 1 - \left(\sqrt{\frac{G(\omega')\mu}{G(\omega)\mu'}}\right)\exp\left(-\frac{\Delta(\omega,\omega')t}{\kappa H}\right),$$

$$\Delta(\omega,\omega') = \sqrt{\frac{1}{\mu^2}+\frac{1}{\mu'^2}+2\frac{(\omega,\omega')}{|\mu\mu'|}}.$$

Here κ is the mean length of phytoelement chords. For $\omega = -\omega'$, i.e. for the exactly backward reflection, attenuation is absent and $\sigma_\kappa = 0$. Thus, modelling the photon free path we use the new attenuation coefficient σ_κ instead of $G(\omega)$. Here κ is the mean linear dimension of a leaf (Kuusk, 1985).

6.5. STATEMENT OF THE INVERSE PROBLEM AND AN ALGORITHM OF ITS SOLUTION

A monodirectional solar radiation flux is incident on the upper surface of a plane canopy layer. Detectors to measure the canopy spectral radiance are positioned on the upper layer boundary. These detectors are designed to obtain three optical parameters of the layer r_L, t_L, r_s and four geometrical parameters b, c, κ and H. Among the above-mentioned parameters the geometrical ones are more important, since they are more difficult to measure.

We suppose that:

1) the detector is placed not high above the upper layer boundary;

2) the atmospheric optical density is negligible compared to the vegetative layer optical density. Therefore, we may neglect the atmospheric scattering of light within the vegetative layer and on its way from the upper layer boundary to the radiation detector. To take into account the influence of the atmosphere on the incident solar radiation flux, the diffuse component should be added to the monodirectional flux. As the addition by no means affects the algorithm of the problem solution, this effect was not taken into account in the first experiments with the model problem.

The desired parameters enter the intensity transfer equation in a complicated way. Therefore, to obtain closed analytical equations for them is difficult or impossible. In addition, these attempts should certainly lead to different equations for different parameters. Using some rather universal algorithm is more preferable. We will consider a modification of the Newton method. Let $\alpha_1^*, \ldots, \alpha_n^*$ be the desired values of the parameters $\alpha_1, \ldots, \alpha_n$ to be identified. Let us denote by

$$R_k \equiv R_k(\alpha_1, \ldots, \alpha_n), \quad k = 1, \ldots, N > n,$$

the values of the spectral radiance in the viewing direction ω_k^* for a vegetation with the parameters $\alpha_1, \ldots, \alpha_n$, respectively. Let

$$R_k^* \equiv R_k(\alpha_1^*, \ldots, \alpha_n^*), \quad k = 1, \ldots, N,$$

be the spectral radiance values measured. To determine the values of $\alpha_1^*, \ldots, \alpha_n^*$ we will consider a system of nonlinear equations

$$R_k(\alpha_1, \ldots, \alpha_n) = R_k^*, \quad k = 1, \ldots, N, \tag{6.12}$$

where R_k are the functions of quite complicated structure. Solving the system by the Newton – Kantorovich method results in iterations

$$\alpha_i^{l+1} = \alpha_i^l + \delta_i^l, \quad i = 1, \dots, n \quad l = 0, 1, \dots, \tag{6.13}$$

where δ_i^l satisfy the system of linear equations

$$\sum_{i=1}^{n} \frac{\partial R_k}{\partial \alpha_i} \delta_i^l = R_k^* - R_k(\alpha_1, \dots, \alpha_n), \quad k = 1, \dots, N.$$

As a zero approximation to the desired parameters we take the prognostic values $\alpha_1^0, \dots, \alpha_n^0$. The process (13) continues until the inequality

$$|R_k^* - R_k(\alpha_1, \dots, \alpha_n)| < \varepsilon$$

holds, where ε is a given small number. Generally speaking, the system (12) is overdetermined. It may be solved by the least-square method. If the system is ill-posed, one may use a regularization. The spectral radiance values and their derivatives are evaluated by the Monte Carlo method (see Section 6).

Sometimes it is convenient to use a modification of the scheme described. Let us represent the quantity R as a sum

$$R = R_1 + (R - R_1),$$

where R_1 is the spectral radiance of the single-scattered photons. Note that in practically important cases (e. g. in the photosynthetically active region of the spectrum) the first-order scattered radiation flux substantially contributes to the total flux. It is easy to check that this statement holds for the corresponding derivatives. Hence, it is reasonable to substitute the derivatives $\partial R_1/\partial \alpha$ in the system (12) for the derivatives $\partial R/\partial \alpha$. The advantage of such a substitution concerns the derivative calculation algorithm (Section 6). The derivatives $\partial R_1/\partial \alpha$ are evaluated faster and more accurately than the derivatives $\partial R/\partial \alpha$. Of course, this substitution decreases to some extent the theoretical convergence rate. Practically, however, this fact may be compensated for by higher accuracy of the evaluation of the derivatives $\partial R_1/\partial \alpha$. In addition, the relative difference of the derivatives $\partial R/\partial \alpha$ and $\partial R_1/\partial \alpha$ is small for small leaf surface index values.

6.6. EVALUATION OF DERIVATIVES BY THE MONTE CARLO METHOD

Let us write the Neumann series (10) in a short form:

$$R = \int Q_0\psi_0 + \int Q_0 K_{01}\psi_1 + Q_0 K_{01} K_{12}\psi_2 + \dots . \qquad (6.14)$$

Here

$$Q_0 = Q(x_0), \quad \psi_i = \psi(x_i), \quad K_{ij} = K(x_i \to x_j).$$

To evaluate the derivative $\partial R/\partial\alpha$ let us denote $\partial f/\partial\alpha = f'$ and differentiate the series (14):

$$\begin{aligned} R &= \int [Q_0'\psi_0 + Q_0\psi_0'] + \int [Q_0' K_{01}\psi_1 + Q_0 K_{01}'\psi_1 + Q_0 K_{01}\psi_1'] + \dots \\ &= \int Q_0'\psi_0 \cdot \left[\frac{Q_0'}{Q_0} + \frac{\psi_0'}{\psi_0}\right] + \int Q_0 K_{01}\psi_1 \left[\frac{Q_0'}{Q_0} + \frac{K_{01}'}{K_{01}} + \frac{\psi_1'}{\psi_1}\right] + \dots \\ &= \int Q_0'\psi_0 \cdot W_0 + \int Q_0 K_{01}\psi_1 W_{01} + \dots . \end{aligned}$$

The latter representation corresponds to the following evaluation algorithm: as before, we simulate the Markov chains of intersections and at every collision point x add the term $\psi(x) \cdot W$ to the estimate $\partial R/\partial\alpha$. Thus, the derivatives and the spectral brightness coefficient are evaluated along the same trajectories. Strict substantiation of this evaluation algorithm is given in Mikhailov (1967). Obviously, the arithmetic cost of a derivative evaluation in comparison with that of the spectral radiance is determined by the cost of evaluating the factor $W_{0\dots i}$.

It has been already mentioned that the calculation of all the quantities $W_{0\dots i}$ with satisfactory precision requires considerable computer time. Sometimes the problem can be facilitated by evaluating only the derivative of the first term in (14). Let

$$R_1 = \int Q_0\psi_0.$$

Then to evaluate the derivative $\partial R/\partial\alpha$ it is sufficient to evaluate the factor

$$W = \frac{Q_0'}{Q_0} + \frac{\psi_0'}{\psi_0}.$$

Let us write down an expression for the quantity R_1. Two kinds of collisions exist, namely the collisions at the foliage surface and those at the soil surface. Accordingly,

$$R_1 = R_{1L} + R_{1s}.$$

Here the first term refers to the part of the radiation, which is single-scattered from the foliage surface, and the second term to the part of the radiation, which is scattered from the soil surface. From (10)–(12) we obtain

$$R_{1L}(\omega^*) = \pi \cdot \frac{G(\omega_\odot)}{|\mu_\odot|} \cdot \frac{P(\omega_\odot \to \omega^*)}{\mu^*}$$
$$\times \int_0^H \exp\left(-\frac{G(\omega_\odot)}{|\mu_\odot|}t - \frac{1}{\mu^*}\int_0^H \sigma_\kappa(t',\omega,\omega^*)dt'\right) dt,$$

$$R_{1s}(\omega^*) = r_s G(\omega_\odot) \exp\left(-\frac{G(\omega_\odot)}{|\mu_\odot|}H - \frac{1}{\mu^*}\int_0^H \sigma_\kappa(t',\omega,\omega^*)dt'\right).$$

Now let us consider the statistical estimates $\partial R_1/\partial\alpha$ of the specific parameters α.

1) The underlying surface albedo r_s. Obviously, r_s depends only on R_{1s}, thus $W_0 = 1/r_s$ and

$$\frac{\partial R_1}{\partial r_s} = \int Q_0\psi_0 W_0 = G(\omega_\odot) \exp\left(-\frac{G(\omega_\odot)}{|\mu_\odot|}H - \frac{1}{\mu^*}\int_0^H \sigma_\kappa(t',\omega,\omega^*)dt'\right).$$

2) The leaf transmittance and reflectance factors r_L and t_L. These parameters affect only the leaf reflectance indicatrix f, which enters the formula determining $P(\omega \to \omega^*)$. Therefore,

$$W_0 = \frac{\psi'_\alpha}{\psi} = \frac{f'_\alpha(\omega_\odot \to \omega^*;\, \omega_L)}{f(\omega_\odot \to \omega^*;\, \omega_L)},$$

$$W_{0r} = \begin{cases} \dfrac{1}{r_L}, & (\omega,\omega_L)\cdot(\omega',\omega_L) > 0, \\ 0, & (\omega,\omega_L)\cdot(\omega',\omega_L) < 0, \end{cases}$$

$$W_{0t} = \begin{cases} 0, & (\omega,\omega_L)\cdot(\omega',\omega_L) > 0, \\ \dfrac{1}{t_L}, & (\omega,\omega_L)\cdot(\omega',\omega_L) < 0. \end{cases}$$

3) The parameters b and c characterizing the leaf normals orientation distribution density. Both functions ψ and Q depend on these parameters:

$$W_{0\alpha} = \frac{Q'_\alpha}{Q} + \frac{\psi'_\alpha}{\psi}, \qquad \frac{Q'_\alpha}{Q} = \frac{G'_\alpha}{G} - \frac{t}{\mu}G'_\alpha,$$

$$\frac{\psi'_\alpha}{\psi} = \frac{P'_\alpha(\omega_\odot \to \omega^*)}{P(\omega_\odot \to \omega^*)} = \int_0^t \left[-\frac{1}{\mu^* \sigma_\kappa(t', \omega, \omega^*)} \right]'_\alpha dt'.$$

Here the integral is calculated analytically.

4) The leaf area index H. From the formula (17) one can see that H affects only the upper integration limit and the expression for σ_κ. Differentiating R_{1L} with respect to the parameter H, we derive the following expression:

$$\frac{\partial R_{1H}}{\partial H} = \pi \frac{G(\omega_\odot)}{|\mu_\odot|} \frac{P(\omega_\odot \to \omega^*)}{\mu^*} \exp\left(-\frac{G(\omega_\odot)}{|\mu_\odot|} H - \frac{1}{\mu^*} \int_0^H \sigma_\kappa(t', \omega, \omega^*) dt' \right)$$

$$- \frac{\partial}{\partial H} \left(\frac{1}{\mu^*} \int_0^H \sigma_\kappa(t', \omega, \omega^*) dt' \right).$$

Thus, the derivative of R_{1L} with respect to H equals the sum of two summands. The augend is a given analytical expression, while the addend is the expectation of the statistical estimate R_{1L}, multiplied by the weighting factor

$$W = \frac{\partial}{\partial H} \left(- \int_0^t \frac{\sigma_\kappa}{\mu^*} \right).$$

Similarly,

$$\frac{\partial R_{1s}}{\partial H} = \frac{G(\omega_\odot)}{|\mu_\odot| + \dfrac{\sigma_\kappa(t', \omega, \omega^*)}{\mu^*}} + \int_0^H \frac{\partial}{\partial H} \left(\frac{\sigma_\kappa}{\mu^*} \right) dt'.$$

5) The parameter κ characterizing the leaf dimensions. It affects only the contribution function ψ:

$$W = \frac{\psi'_\alpha}{\psi} = - \int_0^H \frac{\partial}{\partial \kappa} \sigma_\kappa(t', \omega_\odot, \omega^*) dt'.$$

6.7. ON ESTIMATING THE LEAF-NORMAL DISTRIBUTION FOR A PLANT CANOPY

As it was said above, we use a two-parameter function to approximate the leaf-normal distribution. The proposed method does not allow to identify the parameters b and c with proper accuracy at large values of H ($H \simeq 4$) (see Section 6). Meanwhile, the leaf-normal distribution is one of the most important parameters of the plant canopy.

Therefore, a new approach has been developed to estimate the two parameters b and c of the leaf-normal distribution with good accuracy. Recall that the problem was posed as follows. Let the detector be pointed to the upper surface of the plant layer illuminated by monodirectional solar radiation. The problem is of estimating the unknown coefficients b and c from observed reflectance data (provided the other model parameters are known). The approach is based on a special choice of the view direction, which allows us to obtain simple nonlinear equations for the unknown parameters. The needed parameters can be easily retrieved numerically from those equations.

6.7.1. Iterative method

Here we present an iterative method for solving the inverse problem. In general, the scheme is as follows.

Each iteration is divided into two steps.

Let I^1 and I^2 be some known values of the observed reflected radiation intensity corresponding to two different view angles of the detector and to the values b^* and c^* of the parameters b and c.

At the first step we approximately estimate the values of the single-scattered radiation I_1^1 and I_1^2 (relevant to the values I^1 and I^2).

At the second step we construct a nonlinear system of two equations for estimating the needed parameters b and c and propose an iterative solution to that equation. This completes the iteration. Then we use thus obtained values b and c to perform the next iteration. Now consider the algorithm in detail.

1) Let I be equal to one of the values I_1^1 or I_1^2. We partition I into the sum of I_S and I_M: $I = I_S + I_M$, where I_S gives the value of the single-scattered radiation, I_M is the value of the multiscattered radiation. All these values depend on b and c, so we can write $I_S(b^*, c^*)$ and $I_M(b^*, c^*)$ instead of I_M and I_M, respectively.

In case of a proper detector set (see item 2) below) the value I_S depends on the coefficients b and c far more than I_M does.

Let b_0 and c_0 be some initial approximate values of the unknown coefficients b^* and c^*. Since I_M is less sensitive to the values of b and c, we set approximately:

$$I_M(b^*, c^*) \approx I_M(b_0, c_0).$$

Hence, $I_S(b^*, c^*)$ may be approximately evaluated as:

$$I_S(b^*, c^*) \equiv I(b^*, c^*) - I_M(b^*, c^*) \approx I(b^*, c^*) - I_M(b_0, c_0).$$

Since b_0 and c_0 are known, we are able to calculate $I_M(b_0, c_0)$ by numerically solving the transport equation for the plant canopy. The value $I(b^*, c^*)$ is also known (we recall that $I = I^1$ or $I = I^2$ and both values I^1 and I^2 are known from observations). Thus, at this step we have approximately estimated the values $I_S^{(1)}(b^*, c^*)$ and $I_S^{(2)}(b^*, c^*)$.

2) Now let us consider the process of estimating the coefficients b and c using the known values $I_S^1(b, c)$ and $I_S^2(b, c)$.

Let us choose the view direction ω^* that coincides with the solar radiation direction $\omega_\odot$. Such a choice enables us to simplify the formula for the value of the single-scattered solar radiation. The point is that in the case of such a direction we can get rid of confused expressions in the formula for single-scattered radiation because the extinction coefficient in the view direction will be equal to zero due to the "hot spot" effect (see Section 4), which results in a considerable simplification.

In the general case, we have obtained the following formula for I_S :

$$I_1 = P(\omega_\odot \to -\omega^*) \frac{G(\omega_\odot)}{|\mu_\odot|} \int\limits_0^H d\tau \exp\left(\frac{G(\omega_\odot)}{|\mu_\odot|}\tau\right) \exp(-\sigma_\kappa(t', \omega, \omega^*)\tau). \tag{6.15}$$

Substituting here the formulas for $P(\omega_\odot \to -\omega^*)$ and $\sigma_\kappa(t, \omega_\odot, \omega^*)$ (see Sections 3 and 4), we obtain a rather complicated expression. But in the case $\omega^* = \omega_\odot$ the formula (15) would be simplified owing to the relation $\sigma_\kappa(t', \omega, -\omega) \equiv 0$.

Furthermore, in this case we have:

$$f(\omega^* \to -\omega^*;\ \omega_L) \equiv \frac{1}{\pi} r_L |(\omega_L, \omega^*)| \tag{6.16}$$

Hence,

$$I_1 = \frac{\int\limits_{\Omega+} d\omega_L\, g(\omega_L) |(\omega_L, \omega^*)|^2}{\int\limits_{\Omega+} d\omega_L\, g(\omega_L) |(\omega_L, \omega^*)|} \cdot \frac{r_L/\pi}{|\mu^*|} \cdot \exp\left(-\frac{G(\omega^*)}{|\mu^*|}H\right). \tag{6.17}$$

Below we show that calculation of these two-dimensional integrals in the expression for I_S can be reduced to calculation of one-dimensional integrals that, in turn, can be calculated either by simple formulas or through calculation of series. This follows from the special choice of the view direction.

Taking into account the formula (1), which represents the distribution function g, we can rewrite these integrals as follows:

$$\begin{aligned}\int_{\Omega+} d\omega_L\, g(\omega_L)|(\omega_L,\omega)| &= \int_{\Omega+} d\omega_L\,(1+b\cdot\cos 2\theta_L + c\cdot\cos 4\theta)|(\omega_L,\omega)| \\ &= \int_{\Omega+} d\omega_L\,|(\omega_L,\omega)| + b\cdot\int_{\Omega+} d\omega_L\,\cos 2\theta_L|(\omega_L,\omega)| \\ &\quad + c\cdot\int_{\Omega+} d\omega_L\,\cos 4\theta_L|(\omega_L,\omega)| \\ &= A_1(\omega) + b\cdot B_1(\omega) + c\cdot C_1(\omega),\end{aligned}$$

where

$$A_1(\omega) = \int_{\Omega+} d\omega_L\,|(\omega_L,\omega)|, \qquad B_1(\omega) = \int_{\Omega+} d\omega_L\,\cos 2\theta_L|(\omega_L,\omega)|^2,$$

$$C_1(\omega) = \int_{\Omega+} d\omega_L\,\cos 4\theta|(\omega_L,\omega)|.$$

In exactly the same way

$$\begin{aligned}\int_{\Omega+} d\omega_L\, g(\omega_L)|(\omega_L,\omega)|^2 &= \int_{\Omega+} d\omega_L\,(1+b\cdot\cos 2\theta_L + c\cdot\cos 4\theta)|(\omega_L,\omega)|^2 \\ &= \int_{\Omega+} d\omega_L\,|(\omega_L,\omega)|^2 + b\cdot\int_{\Omega+} d\omega_L\,\cos 2\theta_L|(\omega_L,\omega)|^2 \\ &\quad + c\cdot\int_{\Omega+} d\omega_L\,\cos 4\theta_L|(\omega_L,\omega)|^2 \\ &= A_2(\omega) + b\cdot B_2(\omega) + c\cdot C_2(\omega),\end{aligned}$$

where

$$A_2(\omega) = \int_{\Omega+} d\omega_L\,|(\omega_L,\omega)|^2, \qquad B_2(\omega) = \int_{\Omega+} d\omega_L\,\cos 2\theta_L|(\omega_L,\omega)|^2,$$

$$C_2(\omega) = \int\limits_{\Omega+} d\omega_L \cos 4\theta |(\omega_L, \omega)|^2.$$

The calculation of A, B and C is discussed below.

Thus, for a fixed direction ω, these integrals can be represented as linear forms of b and c, and the expression for $I_S(\omega^*)$ may thus be rewritten as follows:

$$I_S(\omega^*) = E(\omega^*) \frac{A_2(\omega^*) + b \cdot B_2(\omega^*) + c \cdot C_2(\omega^*)}{A_1(\omega^*) + b \cdot B_1(\omega^*) + c \cdot C_1(\omega^*)}, \tag{6.18}$$

which can be regarded as an equation for b and c (provided I_S is known).

Here the value

$$E(\omega) = \frac{r_L/\pi}{|\mu|} \cdot \exp\left(- \frac{G(\omega)}{|\mu|} H \right) \tag{6.19}$$

depends on b and c in a nonlinear way. For two different directions $\omega^{*(1)}$ and $\omega^{*(2)}$ we obtain a system of two nonlinear equations:

$$I_S^1 = E(\omega^{*(1)}) \frac{A_2(\omega^{*(1)}) + b \cdot B_2(\omega^{*(1)}) + c \cdot C_2(\omega^{*(1)})}{A_1(\omega^{*(1)}) + b \cdot B_1(\omega^{*(1)}) + c \cdot C_1(\omega^{*(1)})},$$

$$I_S^2 = E(\omega^{*(2)}) \frac{A_2(\omega^{*(2)}) + b \cdot B_2(\omega^{*(2)}) + c \cdot C_2(\omega^{*(2)})}{A_1(\omega^{*(2)}) + b \cdot B_1(\omega^{*(2)}) + c \cdot C_1(\omega^{*(2)})}.$$

An iterative procedure can be used to solve this system of nonlinear equations.

Let b_0 and c_0 be the initial (and approximate) values of a solution of the system (11). As we have noticed above, $E(\omega)$ depends on b and c. Substituting $b = b_0$, $c = c_0$ into $E(\omega^{*(1)})$ and $E(\omega^{*(2)})$ we calculate the values $E(\omega^{*(1)})$ and $E(\omega^{*(2)})$ and thus reduce the nonlinear system (11) to a linear system for b and c. Solving this linear system we then retrieve new values b_0' and c_0'. With these new values $b = b_0'$ and $c = c_0'$ we repeat the process. Three to four iterations are enough to calculate the solution b_1 and c_1 of the system (11) with accuracy of about 0.5%.

3) With such retrieved values b_1 and c_1, we repeat the whole process restarting from the step 1). Thus we obtain the consequent iterations b_0, c_0; b_1, c_1; $b_2, c_2, \ldots$ for estimating the coefficients b^* and c^*.

6.7.2. On calculating the integrals A, B and C

The calculation of the integrals A, B, and C can be reduced to the calculation of one-dimensional integrals, which, in turn, can be evaluated by means of series expansions.

Consider, for instance, the integral

$$B_1(\omega^*) = \int\limits_{\Omega+} d\omega_L \cos 2\theta_L |(\omega_L, \omega^*)|.$$

First we note that

$$\int\limits_{\Omega+} d\omega_L \cos 2\theta_L |(\omega_L, \omega^*)| = \frac{1}{2} \int\limits_{\Omega} d\omega_L \cos 2\theta_L |(\omega_L, \omega^*)|$$

and then (taking into account that $\cos 2\theta_L = 2\cos^2\theta_L - 1$)

$$\frac{1}{2} \int\limits_{\Omega} d\omega_L \cos 2\theta_L |(\omega_L, \omega^*)| = \frac{1}{2} \int\limits_{\Omega} d\omega_L (2\cos^2\theta_L - 1)|(\omega_L, \omega^*)|.$$

Now we utilize the following formula (see Vladimirov (1981)):

$$\int\limits_{\Omega} d\omega' f(\omega, \omega') g(\omega', \omega'') = 4\pi \sum_{k=0}^{\infty} \frac{2k+1}{4} (f, P_k)(g, P_k) P_k(\omega, \omega''). \qquad (6.20)$$

Here P_k denotes the Legendre polynomials of order k. To reduce the calculation of the integral (12) to (20) we assign the following values in (20):

$$f(x) = 2x^2 - 1, \qquad g(x) = |x|, \qquad \omega = e_3, \quad \omega' = \omega_L, \quad \omega'' = \omega^*,$$

where $e_3 = (0, 0, 1)$. Then in (20)

$$f(\omega, \omega') = 2\cos^2\theta_L - 1 = \cos 2\theta_L, \qquad g(\omega', \omega'') = |(\omega_L, \omega^*)|,$$

$$(\omega, \omega') \equiv (e_3, \omega_L) = \cos\theta_L, \qquad (\omega', \omega'') = (\omega_L, \omega^*), \qquad (\omega, \omega'') = \mu^*.$$

Thus, at a fixed $\mu^* \equiv \omega' \cdot \omega^*$ the calculation of the integral (12)

$$B_1(\omega^*) = \int\limits_{\Omega} d\omega_L \cos 2\theta_L |(\omega_L, \omega^*)| = \int\limits_{\Omega} d\omega' \; f(\omega, \omega') g(\omega', \omega'') \qquad (6.21)$$

may be reduced to calculating the series:

$$4\pi \sum_{k=0}^{\infty} \frac{2k+1}{4} (f, P_k)(g, P_k) P_k(\omega, \omega'')$$

$$= 4\pi \sum_{k=0}^{\infty} \frac{2k+1}{4} \left[\int_{-1}^{+1} dx\, (2x^2 - 1) P_k(x) \right] \left[\int_{-1}^{+1} dx\, |x| P_k(x) \right] \cdot P_k(\mu^*)$$

$$= 4\pi \sum_{l=0}^{\infty} \frac{l+1}{2} \left[\int_{-1}^{+1} dx\, (2x^2 - 1) P_{2l}(x) \right] \left[\int_{-1}^{+1} dx\, x P_{2l}(x) \right] \cdot P_{2l}(\mu^*) = \dots .$$

One can easily notice that all the coefficients A, B, and C depend only on one variable m. The final formulas for A, B, and C are written down below:

$$A_1(\mu) = \frac{1}{2} + \sum_{n=1}^{\infty} \frac{4n+1}{4} \cdot \frac{1}{2} \prod_{k=1}^{n} \left[\frac{2k-1}{2k}\right]^2 \prod_{k=1}^{n} \left[-\frac{2k-1}{2k+4}\right] P_{2k}(\mu),$$

$$B_1(\mu) = -\frac{5}{32} + \sum_{n=2}^{\infty} \frac{4n+1}{4} \cdot \frac{3}{4} \prod_{k=2}^{n} \left[\frac{(2k-3)(2k+1)}{4(k^2-1)}\right]^2 \times \prod_{k=2}^{n} \left[-\frac{2k-1}{2k+4}\right] P_{2k}(\mu),$$

$$C_1(\mu) = -\frac{315}{256} + \sum_{n=3}^{\infty} \frac{4n+1}{4} \cdot \frac{35}{64} \prod_{k=3}^{n} \left[\frac{4k^2-25}{4k^2}\right]^2 \prod_{k=3}^{n} \left[-\frac{2k-1}{2k+4}\right] P_{2k}(\mu),$$

$$A_2(\mu) = \frac{\mu^2+1}{4}, \qquad B_2(\mu) = \frac{3\mu^2-1}{8}, \qquad C_2 \equiv 0.$$

6.8. NUMERICAL RESULTS

In this section we present results of a few model computations of identification parameters of a canopy layer. The general scheme of the calculation is as follows. First, we calculate coefficients of spectral brightness of the transmitted and reflected radiance in several directions for a model medium. Then we use these data for identification of the model parameters. The choice of the observation conditions depends on the parameters to be identified. The location of the detector and the observation direction is chosen so that:

1) the proportion of the once-scattered radiation is as high as possible;

2) the needed parameter (or parameters) could be found using the values of intensity of the once-scattered radiation.

In our calculations the zenith solar angle is $\theta_\odot = 40°$, the observation azimuth angle (except for identification of the parameters b and c) is $\varphi = 0°$.

Below we introduce the detector parameters: its altitude z and zenith observation angle θ.

1) H: $z = 0$; $\theta = 50°$. At these parameter values we observe the transmitted radiation; the angle is close to $\theta_\odot$ ($\theta \neq \theta_\odot$, so that the direct solar radiation does not come to the detector).

2) κ: $z = H$; $\theta = 210°$. At these parameter values we observe the reflected radiation, which is considerably more sensitive to the value κ than the transmitted radiation; the angle θ is close to $\theta_\odot + 180°$ (at $\theta = \theta_\odot + 180°$ the single-scattered radiation does not depend on κ due to the formula for σ_κ).

3) r_s: $z = H$; $\theta = 180°$.

4) r_L, t_L: the two receivers $z_1 = z_2 = H$; $\theta_1 = 230°$, $\varphi_1 = 0°$; $\theta_2 = 0°$, $\varphi_2 = 230°$.

5) b, c.

We used the standard or the modified scheme, as the situation required. At a satisfactory identification, the number of iterations was chosen so as to obtain the first two significant digits of the needed parameter.

First, we considered a plant canopy layer having the leaf area index $H = 4$. The needed parameters of the model and the results of their identification are presented in Table 1. In the first row of the table designations of the needed parameters are presented. In the second and third rows we write down the exact values (to be found) and the initial approximation of those parameters. The fourth row gives the number of iterations of the modified scheme for identification of the needed parameter at the prescribed accuracy. The fifth row gives the number of iterations of the standard scheme. The sixth row gives the relative error of the parameter identification. The parameter κ was easy to identify, so we did not use here the standard scheme. We failed to identify the parameters b and c by means of the standard scheme. That is why Table 1 shows only the number of iterations for the modified scheme.

The same computation has been performed to identify the parameters b and c for $H = 2$. Five iterations of the modified scheme were enough to identify these parameters with an error of about 10%.

Table 1: Results of identification of the parameters of the plant canopy layer

H	κ	r_s	r_L	t_L	b	c
4	0.08	0.2	0.46	0.46	1	1
2	0.04	0.1	0.20	0.20	0	0
5	2	5	3	3	5	5
2	—	2	3	3	—	—
2%	1%	2%	1%	2%	20%	30%

Table 2: Identification of the parameters b and c.

Iteration number	b_n	c_n
0	$b = 0.00$	$c = 0.00$
1	$b = 0.93$	$c = 0.46$
2	$b = 1.00$	$c = 0.95$
*	$b = 1.00$	$c = 1.00$

Iteration number	b_n	c_n
0	$b = 0.00$	$c = 1.00$
1	$b = 1.00$	$c = 0.50$
2	$b = 1.00$	$c = 0.07$
*	$b = 1.00$	$c = 0.00$

Iteration number	b_n	c_n
0	$b = 1.00$	$c = 0.00$
1	$b = 0.23$	$c = 0.61$
2	$b = 0.05$	$c = 0.94$
*	$b = 0.00$	$c = 1.00$

Iteration number	b_n	c_n
0	$b = 1.00$	$b = 1.00$
1	$b = 0.22$	$b = 0.22$
2	$b = 0.06$	$b = 0.06$
*	$b = 0.00$	$b = 0.00$

Thus, and as we have expected, the identification problem can be solved easier at the smaller values of the leaf area index H.

Now we highlight the results of identification of the parameters b and c by the new method and make some remarks concerning the numerical algorithm.

In order to examine the efficiency of the estimation method, different values of the parameters have been considered. Here we present the results for the most difficult cases: $H = 4$; $r_L = t_L = 0.46$; $r_s = 0$; $\kappa = 0.08$.

As unknown values b^* and c^* of the parameters b and c we have considered the following four pairs of numbers: $(1;1)$, $(1;0)$, $(0;1)$, $(0;0)$.

In some sense the listed pairs are "marginal". If the values of the parameters b and c fall outside the interval $[0;1]$, the value of the distribution function g may become negative, which makes no physical sense. The initial approximate values b and c differ at the most from the desired b^* and c^*, namely we take $b = 1 - b^*$, $c = 1 - c^*$. The results are presented in Table 2.

Thus, it is clear that two iterations are enough to satisfactorily estimate the parameters b and c for these example cases.

6.9. SYNOPSIS

The proposed method for solving the integral equation for radiation transfer in a plant canopy by the Monte Carlo method allows to calculate the bidirectional reflection function of the system "earth – plant canopy" for different types of agriculture plants. The method allows to study more complicated models with greater number of parameters.

The computation results agree well with the solution to the integro-differential equation given by the discrete ordinate method.

We conclude with a few remarks on peculiarities of the approach described in this section. To achieve high accuracy in estimating the needed parameters, we need to have high accuracy values for I_M as well. In the examples above, the error in I does not exceed 1%.

The estimation accuracy depends essentially on the accuracy of the approximation for I_1^*. This approximation would be good in case of small reflectance and transmittance coefficients r_L and t_L (at the values $r_L, t_L \leq 0.2$ one iteration is enough).

One of the most serious drawbacks of the proposed approach is that we need to make measurements of the intensity I^* for several sets of sun angles. Refinements of the present approach may be able to overcome this difficulty.

Furthermore, the algorithm might be changed in another way. We can consider a multiparameter density function g using some basic functions other than $\cos 2\theta$ and $\cos 4\theta$. This would extend the estimation procedure to a greater variety of canopy types.

Chapter 7.

X-ray tomography in scattering media

The word *tomography* means “a technique for making a picture of plane section of a three-dimensional body”. Currently this word comes to be used in a wider sense, namely for any technique of reconstructing internal structures in a body from data collected by detectors outside the body.

This problem for different kinds of radiation in use arises in many branches of modern science, see, for example, Herman, Natterer, and Louis (1990).

In X-ray computerized tomography the picture in question is a representation of the distribution of a physical property called X-ray attenuation in some section of the body. In most practical cases, when the scattering may be ignored, this attenuation is closely related to density.

In these cases mathematical problem reduces to a Radon transformation, see Herman (1979), for mathematical background, see Natterer (1986), Smith, Solomon, and Wagner (1977), Smith and Keinert (1985).

Thus, conventional computerized tomography deals with unscattered radiation, and scattered radiation is reckoned as a noise. In the scanning procedure the scattering effect is usually not regarded against the unscattered (straight) radiation effect.

However, in the case of an optically dense medium with high scattering level, the influence of the scattering effect on the body image is too significant to be ignored. So in the case of strong scattering we should process the data obtained through attenuation measurement before we apply the conventional reconstruction procedure.

Many researchers have been interested in image reconstruction of objects hidden in scattering media; Grunbaum, Herman, *et. al.* (1990) developed a discretized approach, some problems with aperture sources have been considered by Anikonov (1985). Some similar problems arise in remote sensing of the Earth's atmosphere, milky liquids, and biological bodies.

We regard radiation as a flux of small particles through a medium, which gives us the possibility to reduce the problem of the scattered radiation propagation to consideration of the 5-dimensional integro-differential transport equation.

Here a new method is proposed, which enables us to separate unscattered and single-scattered radiation from the intensity of the total flux. This method is a significant modification of the method originally suggested by Romanov (1987) to solve inverse problems for hyperbolic equations.

The proposed technique is based on some properties of the radiation field. More specifically, the radiation field induced by a pencil-beam source can be represented as a sum of certain components (see Section 1): a continuous one and distributions concentrated on some lines and surfaces. Actually, consideration of the real medium with a real radiation pencil-beam source reveals two different kinds of jump in the measurement of radiation field at certain points of the phase space.

As the first step, using a special mutual arrangement of the radiation sources and detectors, we measure values of the discontinuities of the first kind, which enables us to estimate the quantity of the unscattered radiation on the spot around x^0 where the detector is located. Then one can retrieve the total extinction coefficient distribution using standard algorithm of computer tomography.

As the next step, we measure values of the jumps of the second kind, which enables us to estimate the quantity of the single-scattered radiation flux at a given point. Then we can retrieve values of the scattering and absorption coefficients at given points of the examined body by solving a system of simple equations.

Here we do not review the Radon transform methods, see Deans (1983), Helgason (1980), and Natterer (1986). Instead we focus on how to reduce the scattering effect on the attenuation measurements and how to partition the extinction coefficient into the scattering and absorption coefficients in particular points of the body.

Another remark: the proposed method has certain limitations. Apparently, in practice the nonscattering signal is difficult to extract by the proposed method in very dense media. The unscattered radiation decays exponentially with respect to the distance, so in case of a relatively large optical distance the nonscattering signal rate reduces to the detector noise rate.

The plan of the chapter is as follows.

In Section 1 the mathematical aspect of the proposed method is described. Section 2 is devoted to the description of a special source-detectors configuration. We also describe here how to get the needed parameters. Section 3 considers some numerical examples. We demonstrate numerical implementation of the proposed method and a simple example of the influence of the scattered radiation on the tomography image. The final Section 4 completes this chapter with discussion.

7.1. MATHEMATICAL BACKGROUND

We use the transport equation to describe the process of γ-rays propagation in the scattering and absorbing media, see Davison (1958).

A phase space (x, t, ω) is 6-dimensional. Three independent coordinates are used to represent position x at the moment t and two others to indicate the direction ω of the particle (ω is a unit vector).

The solution $E(x, t, \omega)$ of the transport equation

$$\frac{1}{v}\frac{\partial E}{\partial t} + \omega \cdot \nabla_x E + \sigma(x)E = \frac{\sigma_s(x)}{4\pi}\int\limits_{S^2} \eta(x, \omega, \omega^*)E(\omega^*)d\omega^* + Q(x, t, \omega)$$

represents the density of the number of particles at the point $x = (x_1, x_2, x_3)$ at a moment t and moving in the direction ω.

Here $Q(x, t, \omega)$ is the source density; $\sigma(x)$ and $\sigma_s(x)$ are total and scattering macroscopic cross-sections (coefficients), respectively:

$$\sigma(x) = \sigma_a(x) + \sigma_s(x), \qquad \sigma(x), \sigma_a(x), \sigma_s(x) \geq 0,$$

$\sigma_a(x)$ is the absorption macroscopic cross-section; $\eta(x, \omega, \omega^*)$ denotes the probability that a particle entering the collision at the direction ω leaves it in the direction ω^*. We call this function the scattering phase function (indicatrix). It is normalized to unity:

$$\frac{1}{4\pi}\int\limits_{S^2} \eta(x, \omega^*, \omega)d\omega^* = 1.$$

Now we give a brief description of the results of investigation of the singular structure of the fundamental solution to the transport equation, which have been originally presented in short communication by Bondarenko (1992).

Let $C_0^\infty(R^4 \times S^2)$ denote a space of C^∞ functions with compact support in $R^4 \times S^2$. We denote the space of test functions (the set $C_0^\infty(R^4 \times S^2)$ with a pseudotopology) by $\mathcal{D}$ and the space of distributions (dual to $\mathcal{D}$) by $\mathcal{D}'$.

Let $\delta(\omega - \omega^0)$ be a delta function on the unit sphere S^2 concentrated at the point $\omega^0 \in S^2$. For any function $\phi(x,t,\omega) \in \mathcal{D}$ we have:

$$\big(\delta(\omega - \omega^0), \phi(x,t,\omega)\big) = \phi(x,t,\omega^0).$$

The distribution $E(x,t,\omega) \in \mathcal{D}'$ satisfying the equation

$$\frac{1}{v}\frac{\partial E}{\partial t} + \omega\cdot\nabla_x E + \sigma(x)E = \frac{\sigma_s(x)}{4\pi}\int\limits_{S^2} \eta(x,\omega,\omega^*)E(\omega^*)d\omega^* + \delta(x-x^0,t)\delta(\omega-\omega^0)$$

and the initial condition

$$E(x,t,\omega)|_{t<0} \equiv 0$$

is called the fundamental solution of the transport equation.

Now let B be the closed cone

$$B = \{x \in R^3, t \in \overline{R^1_+} : vt \geq |x - x^0|\}$$

and K^+ be the region in $R^4 \times S^2$

$$K^+ = B \times S^2.$$

The equation of the line in R^3_x passing through the points $x^0 + vt\omega$ and $x^0 + vt\omega^0$ for fixed $t > 0$, $\omega, \omega^0 \in S^2$ has the form

$$(x - x^0 - vt\omega) \times (\omega - \omega^0) = 0.$$

$L(t,\omega,x^0,\omega^0)$ denotes the closed segment of this line lying between these points. Let us consider two segments in R^3_x: $x = x^0 + v\tau\omega$ and $x = x^0 + v\tau\omega^0$, $\tau \in [0,t]$. We denote by $\Lambda = \Lambda(t,\omega,x^0\omega^0)$ the part of the plane in R^3_x bounded by these segments.

This construction defines a 4-surface Γ^+ in K^+

$$\Gamma^+ = \{x,t,\omega|\{((x - x^0 - vt\omega) \times (\omega - \omega^0)) = 0\} \cap \{K^+\}\}.$$

We also introduce the function

$$\Gamma = \Gamma(x,t,\omega,x^0,\omega^0) = |(x - x^0 - vt\omega^0) \times (\omega - \omega^0)|.$$

Now we define $x^*(x,t,\omega,x^0,\omega^0)$ as the point of intersection of the ray $r = x^0+\omega^0\tau, \tau \in [0,t]$ and the line $r = x^0+\omega\tau$ for any point $x \in \Lambda$. The integral of $\sigma(x)$ over the segment $[x,y]$ in R^3_x (optical distance) is denoted by

$$\rho(x,y) \equiv \exp\left(-|x-y|\int\limits_0^1 \sigma\left(x+s(y-x)\right) ds\right).$$

We also denote by

$$\theta(t) = \begin{cases} 1, & t \geq 0, \\ 0, & t < 0, \end{cases}$$

the Heaviside function.

Theorem 1. *Let $\eta(x,\omega,\omega^*) \in \mathcal{D}(R^3 \times S^2 \times S^2)$, $\sigma(x) \in \mathcal{D}(R^3)$, $\sigma_s(x) \in \mathcal{D}(R^3)$. Then the fundamental solution $E(x,t,\omega)$ can be represented as follows:*

$$E(x,t,\omega;x^0,\omega^0) = \sum_{n=0}^{n=3} S_n(x,t,\omega;x^0,\omega^0)\psi_n + \theta(vt-|x-x^0|)g(x,t,\omega;x^0,\omega^0),$$

where

$$S_0 = \theta(t)\delta(x-x^0-vt\omega)\delta(\omega-\omega^0), \qquad S_1 = \delta(\Gamma^+)$$

$$S_2 = \theta(vt-|x-x^0|)\,\frac{1}{\Gamma}, \qquad S_3 = \frac{\theta(vt-|x-x^0|)}{|\omega-\omega^0|}\ln\left(\frac{ct}{\Gamma}\right),$$

$$\psi_0 = v\rho(x^0,x), \qquad \psi_1 = \frac{v}{4\pi}\rho(x^0,x^*)\rho(x^*,x)\sigma_s(x^*)\eta(x^*,\omega,\omega^0),$$

$$\psi_2 = \varphi(x,t,\omega)\operatorname{arctg}\Phi(x,t,\omega)$$

for some $\Phi(x,t,\omega)$, $c = 4(1+\sqrt{2})$ and functions φ, g, ψ_3 bounded on K^+. Moreover, $\varphi \neq 0$ on $\operatorname{int} K^+$ if

$$\eta(x,\omega,\omega^*) > 0 \quad \text{and} \quad \sigma_s(x) > 0$$

on $K^+ \times S^2$ and K^+, respectively.

The function $\delta = \delta(\Gamma^+)$ is the Dirac delta function concentrated on the 4-surface Γ^+. It affects a test function $\phi(x,t,\omega) \in \mathcal{D}(K^+)$ as follows:

$$(\delta(\Gamma^+),\phi) = \int\limits_{R^1_+} dt \int\limits_{S^2} d\omega \int\limits_{L(t,\omega,x^0,\omega^0)} \frac{dl}{|\omega-\omega^0|}\phi(x,t,\omega).$$

Theorem 2. *Suppose that the conditions of Theorem 1 are satisfied. Then the problem has the unique solution $H(x,\omega)$ in $\mathcal{D}'(R^3 \times S^2)$.*

Further we assume that the indicatrix η depends on the inner product $\mu = (\Omega \cdot \Omega^*)$:

$$\eta(x, \Omega, \Omega^*) \equiv \eta(x, \Omega \cdot \Omega^*) \equiv \eta(x, \mu).$$

Now consider the problem for the time-independent transport equation

$$\Omega \cdot \nabla_x H + \sigma(x) H = \lambda \frac{\sigma_s(x)}{2\pi} \int_{S^2} \eta(x, \Omega \cdot \Omega^*) H(\Omega^*) d\Omega^* + \delta(x - x^0)\delta(\Omega - \Omega^0).$$

In terms of operators we get

$$LH = \lambda S H + \delta(x - x^0)\delta(\Omega - \Omega^0), \quad \lambda = \text{const}.$$

This equation has infinite number of solutions. Below we consider only the main fundamental solution, also called Green's function, which fits the condition of lacking the ingoing radiation in a sufficiently large sphere $S_R \subset R^3_x$:

$$H(x, \Omega; x^0, \Omega^0)|_{x \in S_R, \Omega \cdot N(x) \geq 0} \equiv 0,$$

where $N(x)$ is the inner normal vector at a point $x \in S_R$. Conditions (1) and (3) guarantee the existence of this solution.

Theorem 3. *Suppose that conditions (1) and (3) are satisfied. Then the problem (5), (7) has the unique solution $H(x,\Omega)$ in $\mathcal{D}'(R^3 \times S^2)$.*

The existence theorem is usually proved through construction of the Neumann series

$$H(x, \Omega; x^0, \Omega^0) = \sum_{n \geq 0} \lambda^n H_n(x, \Omega; x^0, \Omega^0)$$

for the integral equation associated with (5), (7)

$$H = \lambda L^{-1} S H + L^{-1} \delta(\Omega - \Omega^0)\delta(x - x^0).$$

Here

$$L^{-1} u(x, \Omega) \equiv \int_0^\infty \exp\left(- \int_x^{x - \Omega t} \sigma(x) dx \right) u(x - \Omega t, \Omega) dt.$$

The representation (4) for the nonstationary equation has been constructed similarly. Thus we directly retrieve the following relation:

$$H_n(x, \Omega; x^0, \Omega^0) = \int_0^\infty S_n(x, t, \Omega; x^0, \Omega^0) \psi_n dt.$$

This technique is known as a descent method. We intend to get a representation for $H(x,\Omega;x^0,\Omega^0)$ akin to (4) using the relation (10) and thereafter to extract the singularities $G_n(x,\Omega;x^0,\Omega^0)$

$$H_n(x,\Omega;x^0,\Omega^0) = G_n(x,\Omega;x^0,\Omega^0)\varphi_n,$$

where $\varphi_n(x,\Omega;x^0,\Omega^0)$ are functions bounded on $R^3 \times S^2$ for any $(x^0,\Omega^0) \in R^3 \times S^2$. We will consider the decomposition of the fundamental solution into the series according to the singularities. Only first two terms of this decomposition are of essential interest to us.

Denote by $\delta(l)$ the δ-function, concentrated on the ray in R^3_x which runs through the point x^0 in the direction Ω^0 for any $\Omega \in S^2$. Then we have, by definition, for any function $\phi(x,\Omega) \in \mathcal{D}(R^3 \times S^2)$

$$(\delta(l), \phi(x,\Omega)) = \int_0^\infty \phi(x^0 + \tau\Omega^0, \Omega)d\tau.$$

For every fixed Ω, x^0, Ω^0, $\Sigma = \Sigma(x^0,\Omega,\Omega^0)$ denotes the part of the plane confined between the rays $r = x^0 + \Omega\tau$ and $r = x^0 + \Omega^0\tau$, $\tau \geq 0$:

$$\{x \in \Sigma \mid x = x^0 + \alpha\Omega^0 + \beta\Omega\}, \quad \alpha, \beta \geq 0.$$

Introduce the distribution $|\Omega \times \Omega^0|^{-1}\delta(\Sigma)$ concentrated on Σ as

$$\left(\frac{1}{|\Omega \times \Omega^0|}\delta(\Sigma), \phi(x,\Omega)\right) = \int_{S^2} \frac{d\Omega}{|\Omega \times \Omega^0|} \int_\Sigma \phi(x,\Omega)d\Sigma.$$

Below we use a distribution $f(x,y)$ of the form $f(x) \cdot 1(y)$, which affects a test function $\phi \in \mathcal{D}(R^{n+m})$ as follows:

$$(f(x)\cdot 1(y), \phi) = \left(f(x), \int \phi(x,y)dy\right) = (1(y)\cdot f(x), \phi) = \int (f(x), \phi(x,y))dy.$$

Such a function is called a function independent of y.

Theorem 4. *Suppose that all the restriction conditions of Theorem 2 are satisfied. Then the main fundamental solution can be represented as follows:*

$$H(x,\omega;x^0,\omega^0) = G_0(x,\omega;x^0,\omega^0)g_0 + G_1(x,\omega;x^0,\omega^0)g_1 + h(x,\omega;x^0,\omega^0),$$

where

$$G_0(x,\omega;x^0,\omega^0) = \delta(l)\delta(\omega - \omega^0), \qquad G_1(x,\omega;x^0,\omega^0) = \frac{1}{|\omega \times \omega^0|}\delta(\Sigma),$$

g_0 and g_1 are bounded functions and $h(x,\omega;x^0,\omega^0) \in L^1_{loc}(R^3 \times S^2)$ for any $x^0 \in R^3$, $\omega^0 \in S^2$.

7.2. THE GENERAL SCHEME OF THE METHOD

The method consists in determining the coefficients of equation (5) in the following order: $\sigma(x)$, $\sigma_s(x)$, $\eta(x, \omega \cdot \omega')$, $\sigma_a(x)$.

7.2.1. Design and arrangement of the detectors

Below we consider two kinds of detectors. We denote them by **D** and **E** (see Figure 10). In fact, we consider coupled receivers $\mathbf{D} = (\mathbf{D_1}, \mathbf{D_2})$ and $\mathbf{E} = (\mathbf{E_1}, \mathbf{E_2})$.

Normally, the tomograph detector $\mathbf{D_1}$ is arranged at a point x^1 against the source of radiation. This detector indicates both unscattered and scattered radiation falling in all directions on a surface element $\delta Q \in R^2$ at the point x^1, where the detector $\mathbf{D_1}$ is located.

In contrast with the conventional scheme, we set the same extra-receiver $\mathbf{D_2}$ at a point $x^1 + \Delta x$ adjacent to the zone of the straight (unscattered) radiation yet out of the zone (see Figure 10). We regard the pair $\mathbf{D} = (\mathbf{D_1}, \mathbf{D_2})$ as the detector of the first kind. Detector of the second kind $\mathbf{E} = (\mathbf{E_1}, \mathbf{E_2})$ is two coupled "monodirectional" receivers.

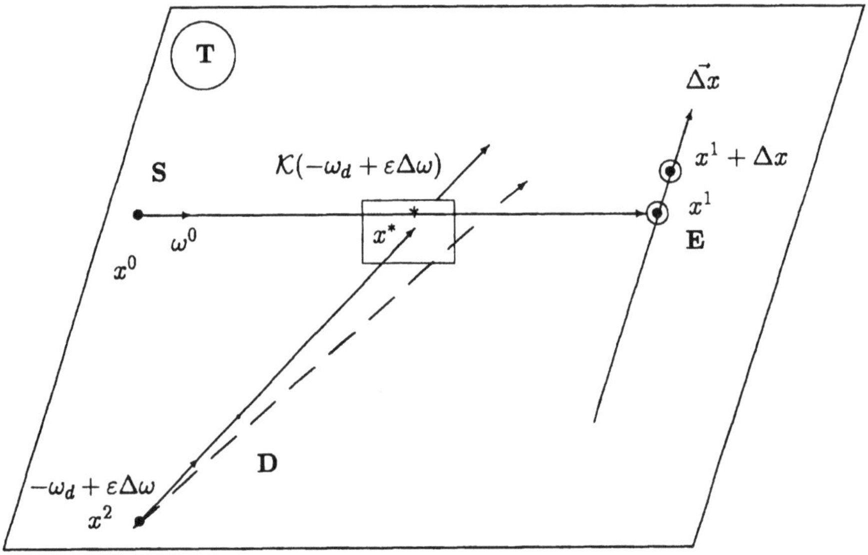

Figure 10: General scheme of the method

Each of these receivers measures the intensity of the radiation incident at the point x_2 in the directions that lie inside the cone $\mathcal{K}(\omega)$ with a view direction ω. Thus, every such receiver is determined by its location point and its view direction.

Fix a point x^* in the source ray to examine. Then choose a view direction ω_d so that the surface of $\mathcal{K}(-\omega_d)$ contains the point x^* and the vector $\Delta\omega$ orthogonal to the plane $\mathbf{T}$ such that the cone $\mathcal{K}(-(\omega_d + \varepsilon\Delta\omega))$ does not intersect the ray $x = x^0 + \tau\omega^0$ for any $\varepsilon > 0$.

Consider now the receiver $\mathbf{E_1}$ with a view direction $\omega_d - \varepsilon\Delta\omega$. The point x^* lies in the intersection of the interior of the cone $\mathcal{K}(-(\omega_d - \varepsilon\Delta\omega))$ and the ray $x = x^0 + \tau\omega^0$ for some $\varepsilon > 0$.

The second receiver $\mathbf{E_2}$ is located at the same point x_2, its view direction being $\omega_d + \varepsilon\Delta\omega$. The cone $\mathcal{K}(-(\omega + \varepsilon\Delta\omega))$ runs in the close vicinity of the point x^*, but it does not intersect the ray $x = x^0 + \tau\omega^0$ for any $\varepsilon > 0$.

7.2.2. Step 1: Reconstruction of the extinction coefficient distribution

In this section we will discuss the method for reconstruction of the distribution of the extinction coefficient.

The first receiver $\mathbf{D_1}$ registers the sum

$$\begin{aligned} I_1(x^1) &= \frac{1}{4\pi} \int\limits_{\delta Q \times S^2} H(x, \omega; x^0, \omega^0) d\omega dQ \\ &= \frac{1}{4\pi} \int_{\delta Q \times S^2} (G_0 g_0 + G_1 g_1 + h) d\omega dQ \\ &= \frac{1}{4\pi} \int_{\delta Q \times S^2} G_0 g_0 d\omega dQ + \frac{1}{4\pi} \int\limits_{\delta Q \times S^2} (G_1 g_1 + h) d\omega dQ \\ &= I_1'(x^1) + I_1''(x^1) \end{aligned}$$

of the unscattered and scattered fluxes at the point x^1 whereas the extra detector $\mathbf{D_2}$ registers only the scattered flux $I_1''(x^1)$ at the point $x^1 + \Delta x$. Since $G_0(x^1 + \Delta x, \omega) = 0$ for any $\omega^0 \in S^2$, we obtain the following expression for the difference between the readings of the two receivers of the first kind:

$$\Delta I_1 = I_1(x^1) - I_1(x^1 + \Delta x) = I_1'(x^1) + (I_1''(x^1) - I_1''(x^1 + \Delta x)).$$

The function

$$F(x) = \int\limits_{S^2} (G_1 g_1 + h) d\omega$$

is integrable, then

$$\lim_{\Delta x\to 0}\Big(I_1''(x^1)-I_1''(x^1+\Delta x)\Big)=0.$$

Hence,

$$\begin{aligned}\delta I_1 &= \lim_{\Delta x\to +0}\left(I_1(x^1)-I_1(x^1+\Delta x)\right)\\ &= I_1'(x^1)+\lim_{\Delta x\to +0}\left(I_1''(x^1)-I_1''(x^1+\Delta x)\right),\end{aligned}$$

$$I_1'(x^1)=\frac{1}{4\pi}\int\limits_{\delta Q\times S^2} G_0 g_0\,d\omega dQ=\frac{1}{\sin\beta}\exp\Bigg(-\int\limits_{x^0}^{x^1}(\sigma_s(x)+\sigma_a(x))dx\Bigg),$$

$$\beta=\angle(\omega^0,\vec{\Delta x})$$

Now, through the measurements we obtain the value $I_1(x^1)$ for $\mathbf{D}_1$ and the value $I_2(x^1+\Delta x)$ of the total flux for $\mathbf{D}_2$. According to the above considerations,

$$\begin{aligned}\Delta I_1 &\equiv (I_1(x^0)-I_1(x^0+\Delta x))\\ &\approx \frac{1}{\sin\beta}\exp\Bigg(-\int\limits_{x^0}^{x^1}(\sigma_s(x)+\sigma_a(x))dx\Bigg),\qquad \Delta x\to +0.\end{aligned}$$

Thus we obtain the value of the attenuated nonscattered radiation at each point x^1. Then the Radon transform procedure can be applied to these data processed that way. Following the conventional scheme of the computed tomography we will be able to reconstruct the distribution of the total extinction coefficient

$$\sigma(x)=\sigma_s(x)+\sigma_a(x)$$

in the examined domain.

7.2.3. Step 2: Reconstruction of the scattering coefficients

Now we will consider how to separate the scattering coefficient and the absorption coefficient at a specific point. It is essential in the proposed scheme that the difference between the readings of the receivers $\mathbf{E}_1$ and $\mathbf{E}_2$ has a jump.

Now we calculate the value of this jump.

Denote by $\mathcal{K}(\omega)$ the open cone with a square section and the direction vector ω. Choose the vector ω_d such that the intersection of the surface of the cone $\mathcal{K}(-\omega_d)$ with the ray $x = x^0 + \tau\omega^0$ contains the interval $(x^* + \Delta x_1, x^* - \Delta x_2)$.

This interval is visible from the point x_2 at the angle $\Delta\gamma$. Denote by $\mathbf{T}^*$ the plane orthogonal to $\mathbf{T}$ passing through points the x^*, x^2.

Choose also the unit vector $-\Delta\omega$ orthogonal to the plane $\mathbf{T}$ that is directed beneath the plane $\mathbf{T}$ on Figure 10.

Note that $G_1(x^2, \omega^*) = 0$ for any $\omega^* \in \mathcal{K}(\omega_d + \varepsilon\Delta\omega)$ and $\varepsilon > 0$. Hence, the flux received by the detector $\mathbf{E}_1$ is equal to

$$I_2(x^2, \omega_d + \varepsilon\Delta\omega) = \frac{1}{4\pi} \int\limits_{\mathcal{K}(\omega_d+\varepsilon\Delta\omega)} \big(G_1(x^2,\omega)g_1 + h(x^2,\omega)\big)\, d\omega = \frac{1}{4\pi} \int\limits_{\mathcal{K}(\omega_d+\varepsilon\Delta\omega)} h(x^2,\omega) d\omega.$$

For any fixed $x \notin \{x = x^0 + \omega^0\tau\}$ the function $h(x^2, \omega)$ is integrable on S^2 and therefore

$$\int\limits_{\mathcal{K}(\omega_d-\varepsilon\Delta\omega)} h(x^2,\omega) d\omega \approx \int\limits_{\mathcal{K}(\omega_d+\varepsilon\Delta\omega)} h(x^2,\omega) d\omega, \quad \varepsilon \to +0.$$

Hence, we obtain the following expression for the difference between the readings of the two receivers of the second kind:

$$\begin{aligned}\Delta I_2 &= I_2(x^2, \omega_d - \varepsilon\Delta\omega) - I_2(x^2, \omega_d + \varepsilon\Delta\omega)\\ &= \frac{1}{4\pi}\Bigg(\int\limits_{\mathcal{K}(\omega_d-\varepsilon\Delta\omega)} G_1(x^2,\omega) g_1 d\omega + \int\limits_{\mathcal{K}(\omega_d-\varepsilon\Delta\omega)} h(x^2,\omega) d\omega \\ &\quad - \int\limits_{\mathcal{K}(\omega_d+\varepsilon\Delta\omega)} h(x^2,\omega) d\omega \Bigg)\\ &\approx \frac{1}{4\pi} \int\limits_{\mathcal{K}(\omega_d-\varepsilon\Delta\omega)} G_1(x^2,\omega) g_1 d\omega, \qquad \varepsilon \to +0.\end{aligned}$$

Denote

$$\psi(x) \equiv \frac{1}{4\pi} \int\limits_{\mathcal{K}(\omega_d-\varepsilon\Delta\omega)} G_1(x^2,\omega) g_1 d\omega = \frac{1}{4\pi} \int\limits_{\mathcal{K}(\omega_d-\varepsilon\Delta\omega)} \frac{g_1\delta(\Sigma)}{|\omega \times \omega^0|} d\omega.$$

Then for $\eta(x) \in \mathcal{D}(x)$

$$\mathcal{I} \equiv (\psi(x), \eta(x))_{R^3} = \frac{1}{4\pi} \int\limits_{\mathcal{K}(\omega_d - \varepsilon\Delta\omega)} \frac{d\omega}{|\omega \times \omega^0|} \int\limits_{R^3} \delta(\Sigma) g_1 \eta(x) dx$$
$$= \frac{1}{4\pi} \int\limits_{\mathcal{K}(\omega_d - \varepsilon\Delta\omega)} \frac{1}{|\omega \times \omega^0|} \int\limits_{\Sigma} g_1 \eta(x) d\sigma d\omega$$
$$= \frac{1}{4\pi} \int\limits_{\mathcal{K}(\omega_d - \varepsilon\Delta\omega)} \int\limits_{\Sigma} \frac{g_1 \eta(x) \cos\psi d\sigma d\psi d\varphi}{|\omega \times \omega^0|}.$$

The element dV of R^3_x is equal to

$$dV = \frac{|(x - x^0) \times \omega^0|}{|\omega \times \omega^0|} d\varphi d\sigma$$

Hence,

$$\mathcal{I} \approx \frac{1}{4\pi} \int\limits_{\Delta V} \frac{g_1 \eta \Delta\gamma dV}{|(x - x^0) \times \omega^0|}, \qquad \Delta\gamma \to +0.$$

Here ΔV is the domain in R^3_x confined between the two planes $\Sigma(x^0, \omega_d, \omega^0)$ and $\Sigma(x^0, \omega^*, \omega^0)$, where $\omega^* \in \mathbf{T}^*$ and the angle between $\mathbf{T}$ and ω^* is equal to $\Delta\gamma$.

Finally, we have

$$\Delta I_2 \approx \frac{\sigma_s(x^*)\rho(0, x^*)\rho(x^*, x)\eta(x^*, \hat{\omega} \cdot \omega^0)}{4\pi\varrho(x^2, x^0, \omega^0)} \Delta\gamma, \qquad \Delta\gamma \to +0, \quad \varepsilon \to +0.$$

Here $\varrho(x^2, x^0, \omega^0) = |(x^2 - x^0) \times \omega^0|$ is the distance between the point x^2 and the line passing trough the point x^0 in the direction of ω^0 and

$$\hat{\omega} \equiv \frac{x^2 - x^*}{|x^2 - x^*|}.$$

It is possible to derive the exact expression for the value of the jump in the case of circular cone. Let ϑ be the angle between the plane $\mathbf{T}$ and the direction vector of the scanning cone of the detector. This angle is equal to ϑ_0 if the cone touches the plane $\mathbf{T}$.

It follows from the calculation similar to the previous one that the derivative $dI_2/d\vartheta$ has a jump at $\vartheta = \vartheta_0$. Its value is

$$\delta\left(\frac{dI_2(\vartheta)}{d\vartheta}\right) = \frac{\sigma_s(x^*)\rho(0, x^*)\rho(x^*, x)\eta(x^*, \hat{\omega} \cdot \omega^0)}{4\pi\varrho(x^2, x^0, \omega^0)}.$$

Since we have already recovered the distribution of the coefficient $\sigma(x)$, the values $\rho(0, x^*)$ and $\rho(x^*, x)$ are known. Thus, all the parameters in this formula are known except $\sigma_s(x^*)$ and $\eta(x^*, \hat{\omega}\cdot\omega^0)$. We would have no trouble with this term in the case of isotropic scattering (then $\eta(x^*, \omega\cdot\omega^0) \equiv 1$). But in the general case we have to estimate this value before calculating $\sigma_s(x)$.

Consider a piecewise approximation of the function $\eta(x^*, \mu)$. Let μ_i such that

$$-1 = \mu_1 < \mu_2 < ... < \mu_n < +1, \quad \mu_i = \cos\varphi_i \equiv (\omega_{d_i} \cdot \omega^0),$$

be the points of the approximation, and p_i be the data obtained in measurements.

We set up n detectors in such positions that the angles between the source and the detector directions are equal to $\varphi_1, \varphi_2, \dots, \varphi_n$.

Thus we obtain n simple equations of the form

$$\sigma_s(x^*)\eta(x^*, \mu_i) = p_i, \quad i = 1, \dots, n.$$

These are the equations for the unknown $\sigma_s(x^*)$ and $\eta(x^*, \mu_i)$. The function $\eta(x^*, \mu)$ is normalized by the condition

$$1 = \int_{-1}^{+1} \eta(x^*, \mu)d\mu \equiv \sum_{i=1}^{n} \eta(x^*, \mu_i)(\mu_{i+1} - \mu_i), \quad \mu_{n+1} \equiv 1.$$

This condition can be regarded as an extra linear equation for the unknown $\sigma_s(x)$ and $\eta(x^*, \mu_i)$. Now $\sigma_s(x)$ can be easily found from $n + 1$ equations (as well as the values $\eta(x^*, \mu_i)$, if required).

7.2.4. Step 3: Reconstruction of the absorption coefficient distribution

Finally, as long as the absorption coefficient $\sigma_a(x) = \sigma(x) - \sigma_s(x)$, we can find it at the same time as $\sigma_s(x)$.

7.3. NUMERICAL RESULTS

Our investigations are not directly concerned with the reconstruction technique. Numerical experiments have been carried out to examine the efficiency of numerical implementation of the proposed method and to display the scattering effect on the body image reconstruction. Here we confine our account to a purely qualitative discussion.

The object to investigate is a sphere D of radius 1 divided into two distinct regions: a concentric sphere of radius 0.5 and the rest. The first subdomain is filled with scattering nonabsorbing substance. On the contrary, the second subdomain is filled with absorbing nonscattering substance. The scattering and absorption coefficients σ_s and σ_a are constant over the correspondent subdomains and equal to each other: $\sigma_s = \sigma_a = 10$. Thus, the value of the total extinction coefficient σ is constant over the whole larger sphere: $\sigma = 10$.

The data has been obtained through simulation of 3D-process of the radiation propagation by the Monte Carlo method.

7.3.1. Computing the jumps of the first and the second kind

Our prime interest here was with the accuracy of estimating the flux of unscattered and once-scattered radiation. We calculated these two values through the values of the jumps of the first and the second kind, respectively.

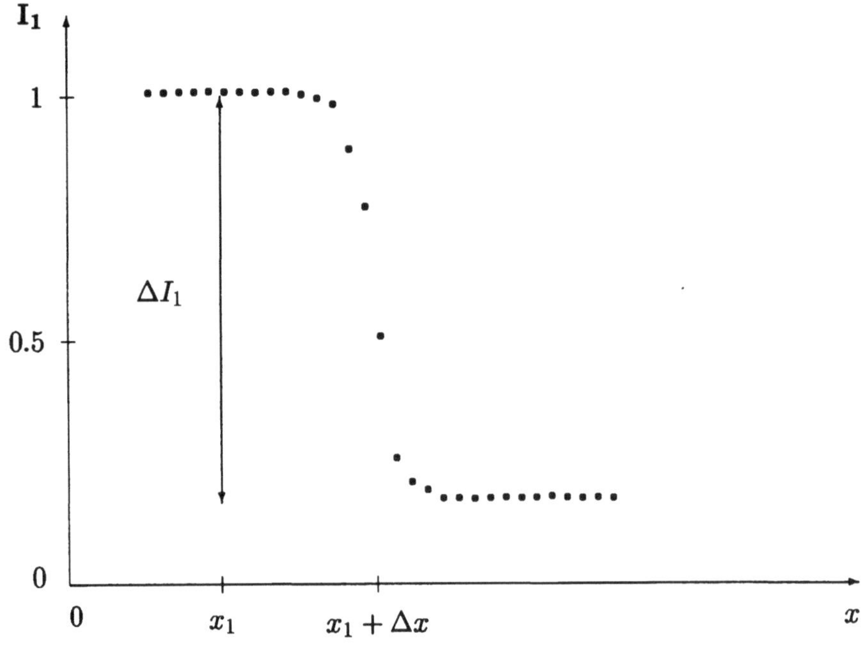

Figure 11: Jump of the first kind

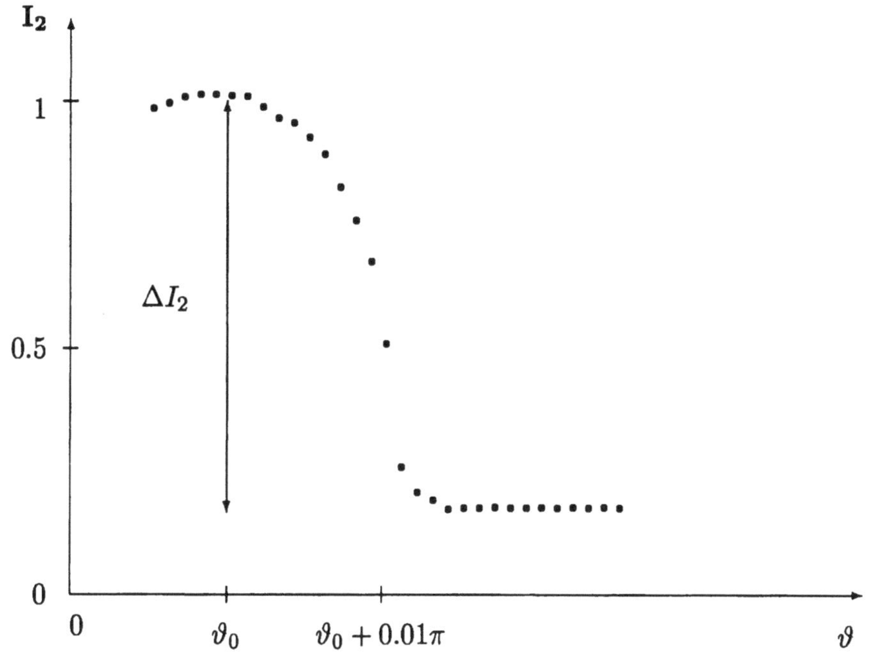

Figure 12: Jump of the second kind

As we said before, the data has been obtained by the Monte Carlo method of computation. One of the advantages of this method is that it enables one to decompose the amount of the scattered radiation into series of amounts with respect to the number of the scattering act.

The curve in Figure 11 shows the dependence between the intensity of the total flux coming into the view of the receiver of the first kind and the distance x.

The curve in Figure 12 shows the dependence between the intensity of the scattered flux coming into the view of the receiver of the second kind (the case of circular cone) and the rotation angle ϑ for the vision direction ω.

One can clearly see the "jumping difference" in the readings of the detector. The errors of estimating the values of unscattered and once-scattered radiation were less then 1 %. Thus, in the case of a real problem these estimations would depend on technical feasibility and the precision of the measurements rather than on the precision of the method.

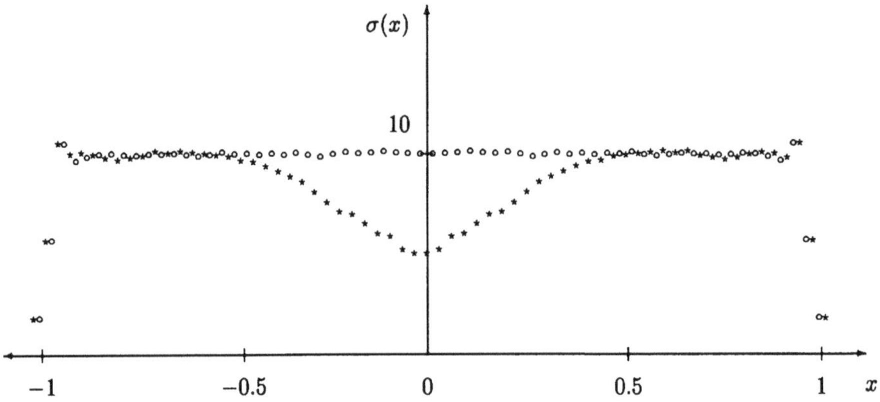

Figure 13: Cross-section for the restored $\sigma(x) \equiv 10$; $\circ$ — homogeneous media, $\star$ — with scattering inhomogeneity

7.3.2. Scattering effect on body image reconstruction

In the previous section we have shown the way to process the data before the Radon transform. Here we apply the Radon transform to unprocessed data in order to show a distortion effect caused by the scattered radiation.

Here the contribution of the scattered flux to the total one ranges up to 40 % owing to the large value of the scattering coefficient ($\sigma_s = 10$).

We reconstructed the image of the central cross-section of the sphere. Figure 13 represents the cross-section of that image along the diameter.

One can see that the recovered coefficient in the inner domain slopes down around the center of the sphere.

This is the distortion in the image caused by the scattering effect. It is natural to expect the same effect in the case of an arbitrary domain filled with scattering substance.

7.4. CONCLUSIONS

In this chapter we present a new approach for data processing in the case of high rate of scattered radiation. Now we summarize its main distinctive points.

Consider again the domain D described above in Section 4. We can process the data retrieved by scanning in the following ways:

1. Standard ST without correction of scattering effects leads to strong distortion in the image. In this case we get the plot [$\star$] in Figure 13.

2. There are a few methods (including the described one) that allow to separate the unscattered part of the radiation. They allow to get the distribution of the total coefficient $\sigma(x)$ over the whole examined body. In this case we get the plot [$\circ$] in Figure 13.

3. The proposed method allows, in addition to mentioned above, to calculate values of the scattering coefficient at any given points. It does not require information obtained from the scanning of the whole body, which consequently leads to smaller waste of the computer time for processing the data. Maybe one can consider this fact as an advantage of the proposed approach. Apparently, an additional information about the distribution of the scattering coefficient might be useful in medical investigation (in discovering "scattering" inhomogeneities, for instance).

4. The proposed method might be considered as an initial step to the final solution of the problem of tomography with scattering. Obviously, technical obstacles put certain limitations to its application. There exists a problem of pencil-beam source. And if we employ laser as a source, a problem of strong attenuation in biological media arises. Apparently, it is not easy to arrange the proposed source-detector configuration. These are the problems to be solved. It is up to practical specialists to take a final decision on the applicability of the method.

Appendix A.

Integral equation of the second kind with stochastic kernel

In this appendix we propose a method for solving integral equation of the second kind with stochastic kernel by the Monte Carlo method. Such equations arise in solving differential equations.

A.1. INTRODUCTION

We consider the integral equation of the second kind

$$f = \lambda K f + \varphi, \tag{A.1}$$

where K is a completely continuous linear operator. Let the operator $K : H \to H$ act in a separable Hilbert space H. The spectrum of the operator K is discrete.

Let r_0 be the eigenvalue of the operator K with maximum absolute value, and let $\lambda = 1/r_0$.

Without loss of generality, we may assume that its multiplicity is equal to unity. Let k_0 be the corresponding eigenfunction.

If the inequality $|\lambda| < \lambda_0$ holds, the solution of equation (1) can be presented in the form of the Neumann series

$$f = \varphi + \lambda K \varphi + \ldots + (\lambda K \varphi)^n + \ldots$$

and the linear functionals

$$I = (f, h) \equiv \int f h$$

can be represented in the form of the series

$$(f,h) = (\varphi,h) + (\lambda K\varphi,h) + \ldots + ((\lambda K\varphi)^n,h) + \ldots$$

and then computed by a conventional procedure of the Monte Carlo method. We should model Markov chains $x_0, x_1, x_2, \ldots$ according to the initial and transition probability densities φ and $k(x_{k-1}, x_k)$, respectively (here k is the kernel of the operator K). Then, as usual, we calculate the sum of the random series

$$\varphi(x_0) + \varphi(x_1) + \ldots$$

that converges with probability 1 (for details see Spanier and Gelbard (1969)).

However, in the case $|\lambda| = \lambda_0$ this statistical estimate makes up a nonconvergent series that is divergent with probability equal to unity, although the solution of equation (1) exists.

In this appendix a method is presented that makes it possible to overcome this difficulty.

A.2. DECOMPOSITION OF THE SOLUTION

In this section we cite some elements of the theory of Fredholm equations which we will need in further considerations. Now, let $|\lambda| = \lambda_0$. Assume that the eigenfunction k_0 is known. Decompose the space H into the direct sum

$$H = N \oplus L.$$

Here N is a one-dimensional proper subspace of the operator K corresponding to the eigenvector k_0:

$$N = \{z : \ z = tk_0\},$$

and the linear subspace L is orthogonal to (N^*):

$$L = (N^*)^\perp.$$

Here N^* is a one-dimensional subspace corresponding to the eigenfunction k_0^* of the conjugate operator K^* corresponding to the eigenvalue λ_0:

$$N^* = \{z : \ z = tk_0^*\}.$$

Every vector $z \in H$ may be uniquely represented as a sum

$$z = \overline{z} + \widetilde{z}, \quad \overline{z} \in N, \ \widetilde{z} \in L; \tag{A.2}$$

here $\overline{z}$ and $\widetilde{z}$ are the affine projections of the vector z on the linear subspaces N and L, respectively.

It is easy to see that if the equality $(\varphi, k_0^*) = 0$ holds, then we have $(K\varphi, k_0^*) = (\varphi, K^* k_0^*) = (\varphi, k_0^*) = 0$. Hence, if the vector $\varphi \in L$, then $K\varphi \in L$. So, we see that N and L are invariant subspaces of the operator K.

Denote by $\widetilde{K}$ the restriction of the operator K on the invariant subspace L. In accordance with the decomposition (2), equation (1) decomposes into the sum of independent equations

$$\overline{f} = \frac{\lambda}{\lambda_0} \overline{f} + \overline{\varphi}, \tag{A.3}$$

$$\widetilde{f} = \lambda \widetilde{K} \widetilde{f} + \widetilde{\varphi}. \tag{A.4}$$

The solution of equation (3) can be easily retrieved:

1) $\lambda = \lambda_0$. Equation (3) turns into the Fredholm condition for the solvability of equation (1):

$$\overline{\varphi} = 0.$$

With the extra condition $\overline{f} = 0$ the solution of equation (1) is unique:

$$f = \widetilde{f}.$$

2) $\lambda = -\lambda_0$. Then equation (3) takes the form

$$\overline{f} = -\overline{f} + \varphi,$$

which results in

$$\overline{f} = \varphi/2.$$

The unique solution of equation (1) is equal to

$$f = \overline{f} + \frac{\overline{\varphi}}{2}.$$

The operator $\widetilde{K}$ is completely continuous as well. Let $r_1, r_2, \ldots$ be the points of its discrete spectrum grouped in descending order of their absolute values. According to the assumption, the inequality $r_0 > |r_1|$ holds. Hence, the spectral radius of the operator $\lambda\widetilde{K}$ is less than unity when $|\lambda| = 1/r_0$, and the solution of equation (4) (unlike (1)) can be represented as a convergent Neumann series.

We only need to recall that

$$\widetilde{K}\widetilde{\varphi} \equiv K\widetilde{\varphi},$$

hence the solution $\widetilde{f}$ can be estimated by the Monte Carlo method.

The described method can be applied to operators K with stochastic kernel $k(x, y)$:

$$k \geq 0, \qquad \int k(x, y)dy \equiv 1.$$

In this case the eigenvalue $r_0 = 1$, the corresponding eigenfunction $k_0(x) \equiv 1$, and hence $\overline{\varphi}(x) \equiv \overline{\varphi} = \text{const}$.

It is clear that the mapping $\varphi \to \overline{\varphi}$ is a linear functional that can be represented by the element k_0^*:

$$\overline{\varphi} = (\varphi, k_0^*),$$

and k_0^* is normed by the condition $\int k_0^* = 1$.

If the operator K is self-adjoint then $k_0^* \equiv \text{const}$, hence

$$\overline{\varphi} = \int \varphi.$$

But in the general case the function k_0^* is unknown.

A.3. NUMERICAL METHOD FOR SOLVING THE EQUATION

As usual, we wish to find a linear functional $I = (f, p)$ of a solution of equation (1), where f is a solution of the equation

$$f = -K^* f + \varphi.$$

Here K is an integral operator with stochastic kernel. Coming to the adjoint problem, we will find the value

$$(g, \varphi) = (f, p),$$

where g is a solution of the equation

$$g = -K^* g + p.$$

This value can be represented as a sum

$$(g, \varphi) = (g, \overline{\varphi}) + (g, \widetilde{\varphi}). \tag{A.5}$$

First, we consider a method for the calculation of the second term.

1. Let $f = \overline{f} + \widetilde{f}$. Then $\widetilde{f}$ satisfies the equation

$$\widetilde{f} = -K\widetilde{f} + \widetilde{\varphi}, \tag{A.6}$$

and

$$\sum_{n=0}^{\infty}(-1)^n K^n \widetilde{\varphi} = \widetilde{f},$$

$$\sum_{n=0}^{\infty}(-1)^n (K^n \widetilde{\varphi}, p) = (\widetilde{f}, p). \tag{A.7}$$

It is clear that the terms of the series

$$\sum_{n=0}^{\infty}(-1)^n (K^{*n} p, \widetilde{\varphi}) \tag{A.8}$$

coincide with the corresponding terms of the convergent series (7). Hence, the series (8) converges to

$$(g, \widetilde{\varphi}) = (\widetilde{f}, p).$$

This value is suited to calculation by the Monte Carlo method. We model a Markov chain $x_0, x_1, x_2, \dots$, where the point x_0 is distributed according to the probability density $q(x_0)$, and the transition probability density for $x_{n-1} \to x_n$ is equal to $K(x_{n-1}, x_n)$. The contribution of the trajectory $\{x_n : n = 0, 1, \dots\}$ is equal to

$$\frac{p(x_0)}{q(x_0)}\,(\widetilde{\varphi}(x_0) - \widetilde{\varphi}(x_1) + \dots).$$

Since K is a stochastic (not substochastic) operator, here we cannot introduce absorption in a "natural" way. Therefore, we will break the trajectories in an artificial way. We fix a "large enough" integer N and break the trajectories at the N-th step (N is the same for all the trajectories).

Take a note that infinite series which represents the estimate calculated for a trajectory diverges with unit probability. It is only the series made up of the corresponding mathematical expectations that converges.

Write down the estimates for the trajectory contributions

1-st trajectory: $\quad \dfrac{p(x_0^1)}{q(x_0^1)}\,(\widetilde{\varphi}(x_0^1) - \widetilde{\varphi}(x_1^1) + \widetilde{\varphi}(x_2^1) - \dots)$

2-nd trajectory: $\frac{p(x_0^2)}{q(x_0^2)}\,(\widetilde{\varphi}(x_0^2) - \widetilde{\varphi}(x_1^2) + \widetilde{\varphi}(x_2^2) - \dots)$

....

n-th trajectory: $\frac{p(x_0^n)}{q(x_0^n)}\,(\widetilde{\varphi}(x_0^n) - \widetilde{\varphi}(x_1^n) + \widetilde{\varphi}(x_2^n) - \dots)$

In this table each series in a row diverges with unit probability, and each series in a column converges.

To find N large enough to achieve the given accuracy, we should estimate the rate of convergence of the series (8), which is the same as for the series (7). The lower limit for this rate may be taken equal to the rate of convergence of the geometric progression $\sum q^n$, $q = \|\widetilde{K}\|$.

The function $\widetilde{\varphi}$ may be represented as a difference $\widetilde{\varphi} = \varphi - \overline{\varphi}$. The value $\overline{\varphi}$ can be found as a limit:

$$K^n\varphi = \overline{\varphi} + K^n\widetilde{\varphi} \longrightarrow \overline{\varphi}.$$

Hence,

$$M_{tr}\ \varphi(x_N) \longrightarrow \overline{\varphi}\int q = \overline{\varphi}$$

as $N \to \infty$. Here M_{tr} is the designation of mathematical expectation over the trajectories $\{tr : x_0, x_1, x_2, \dots\}$.

The value $\overline{\varphi}$ may be found by preliminary computation: to do this, we may model the trajectories $\{x_n\}$ of length N. However, it is more convenient to act the following way. Consider a contribution ζ to the estimate for $(g, \widetilde{\varphi})$ over the trajectory:

$$\begin{aligned}
\zeta_N &= \frac{p(x_0)}{q(x_0)}\,(\widetilde{\varphi}(x_0) - \widetilde{\varphi}(x_1) + \dots \pm \widetilde{\varphi}(x_N)) \\
&= \frac{p(x_0)}{q(x_0)}\,((\varphi(x_0) - \overline{\varphi}) - (\varphi(x_1) - \overline{\varphi}) + \dots \pm (\varphi(x_N) - \overline{\varphi})) \\
&= \left[\frac{p(x_0)}{q(x_0)}\,(\varphi(x_0) - \varphi(x_1) + \dots \pm \varphi(x_N))\right] - \left[\frac{p(x_0)}{q(x_0)}\ \frac{1 + (-1)^N}{2}\,\overline{\varphi}\right] \\
&= \zeta_N - \overline{\varphi}\eta_N.
\end{aligned}$$

Taking into account that

$$M_{tr}\,\frac{p(x_0)}{q(x_0)} = \int p$$

and

$$\overline{\varphi} \simeq M_{tr}\, \varphi(x_N),$$

we obtain

$$M_{tr}\, \zeta_N = M_{tr}\, [\zeta_N - \eta_N \overline{\varphi}] \simeq M_{tr} \left(\zeta_N - \varphi(x_N) \frac{1 + (-1)^N}{2} \int p \right).$$

Thus, the value $\overline{\varphi}$ may be estimated using the same trajectories as those for $M_{tr}\ \zeta$. The error caused by the choice of the number N will be approximately the same — of the order $\|\widetilde{K}\|^{N+1}$. The estimate over the trajectory has the form

$$\frac{p(x_0)}{q(x_0)} \sum_{n=0}^{N} (-1)^N \varphi(x_n) - \left(\frac{1 + (-1)^n}{2} \int p \right) \varphi(x_N).$$

2. Now we study the estimate for the second term $(g, \overline{\varphi})$:

$$(g, \overline{\varphi}) = \overline{\varphi} \int g = \frac{\overline{\varphi}}{2} \int p$$

in the sum (5). The estimate has the form

$$\varphi(x_N) \cdot \frac{1}{2} \int p.$$

The final estimate for $(g, \varphi) = (g, \overline{\varphi}) + (g, \widetilde{\varphi})$ over the cut-off trajectory $x_0, x_1, \ldots, x_N$ is

$$\frac{p(x_0)}{q(x_0)} \sum_{n=0}^{N} (-1)^N \varphi(x_n) - \left(\frac{(-1)^n}{2} \int p \right) \varphi(x_N).$$

The value of the integral of p can be found by a preliminary computatation.

A.4. NUMERICAL EXAMPLES

It is known that some boundary value problems for differential equations can be reduced to Fredholm integral equations with stochastic kernel. Two numerical problems for the Laplace differential equation have been considered to examine the efficiency of the proposed method.

Example 1. We consider the inner Dirichlet problem:

$$\Delta u = 0, \qquad u|_\Gamma = \varphi.$$

Here Γ is the boundary of a convex region D. The problem is to calculate the value $u(x)$, $x \in D$. This problem is equivalent to the problem of calculating the scalar product (μ, K_x), where μ is a solution to the equation $\mu = -K\mu + \varphi$ (Vladimirov, 1981).

Here K is an operator with stochastic kernel $K(x, y)$, $x, y \in \Gamma$:

$$K(x, y) = \frac{\cos \varphi_{xy}}{2\pi\, |x - y|^2},$$

where φ_{xy} is the angle between the vector $\overrightarrow{xy}$ and the normal to the boundary surface Γ at the point y. The function K_x is defined as

$$K_x(y) = K(x, y), \quad x \in D,\ y \in \Gamma.$$

One can check that the kernel K is actually a stochastic kernel. Therefore, we cannot use the conventional Monte Carlo method for solving the integral equation with kernel K. So, we apply the proposed method.

It is convenient to take the probability density $q(y_0) = K_x(y_0) = p(y_0)$. There is no "weight" in the estimate since $\int p = 1$.

We have taken D as a hemisphere $\{x : |x| \le 1,\ x_3 \ge 0\}$ and the function φ in the form

$$\varphi(x) = \frac{1}{|x|}.$$

We seek the solution at the point $x_0 = (0; 0; 0.5)$. The exact value is equal to

$$u(x_0) = \frac{1}{|x_0|} = \frac{2}{3} \approx 0.66667.$$

The result of the computation is presented in the first row of Table 1.

Table 1: Numerical results

Example	Exact value	Relevant error	Number of trajectories
1	0.6667	0.1 %	15000
2	0.0370	0.8 %	16000

Example 2. The next problem is the outer Dirichlet problem:

$$\Delta u = 0, \qquad \left.\frac{\partial u}{\partial n}\right|_{\Gamma} = \varphi,$$

$$u(x) \to 0, \quad |x| \to \infty.$$

This problem can be reduced to the adjoint integral equaton. We have no "natural" probability density q in this case. In calculation we used a density proportional to the Lebesgue measure. We have taken the domain D as a sphere $\{x : |x| \leq 1\}$ and the function φ in the form $\varphi(x) = x_3$.

The solution is the function

$$u(x) = \frac{1}{|x|^3}.$$

We seek the solution at the point $x_0 = (1; 2; 2)$. The exact value is equal to

$$u(x_0) = \frac{1}{27} \approx 0.0370.$$

The result of the computation is presented in the second row of Table 1.

Appendix B.

Computation of radiation field in optically thick medium by the Monte Carlo method

In this appendix we treat the problem of computing the characteristics of the scattered radiation field in an optically thick medium by the Monte Carlo method. This is done by means of a certain transformation of the transport equation for the radiation transfer. The transformation allows us to obtain a new probability density for modelling trajectories. We give an illustrative description of the method based on a study of particle trajectories in the dense medium and a list of limitations for the use of the method. Finally, results of a numerical experiment are presented.

B.1. INTRODUCTION

Solving of the problem of computing the characteristics of the scattered radiation field faces serious difficulties. Different methods for solving these problems have been developed. They are not universal and there are limitations in their application. We note the following.

Diffusion approximation allows reducing the confused transport equation to a simple differential equation. But this approximation can be only applied in the case of isotropic phase function. And even then this approximation can be regarded as very rough.

Exponential transformation is applied in the Monte Carlo computations. It is based on the use of an approximation for the importance function, which in turn may be obtained from the asymptotical solution of the transport equation in a half-space. According to the sense of the exponential transformation, it can be applied in the case of radiation transfer in a half-space or in a slab layer of large thickness.

The proposed method also has certain limitation in its application. We will discuss it in one of the final sections. The main idea of the method is based on separating the δ-function from the probability density of the free path length. The same idea has been used in Chapter 2.

Computation of the radiation transfer in an optically dense media by the Monte Carlo method involves modelling long photon trajectories. This leads to significant waste of computation time. The proposed method gives an approximation to the transport equation that corresponds to the process with shorter trajectories.

Here the term "short trajectory" means a trajectory with small number of collisions. Modelling of collisions consists of modelling free path lengths between collisions and modelling flight directions between the collision points. This modelling is the main or at least a significant part of the Monte Carlo computation of the radiation transfer characteristics.

B.2. TRANSFORMATION OF THE TRANSPORT EQUATION

We use the following notation:

$X = R \times \Omega$ is the phase space of the equation;

$x = (r, \omega)$ are generic points of the phase space;

R is the geometric space of collision points $r = (x, y, z)$ with Cartesian coordinates;

Ω is the space of unit vectors $\omega = (a, b, c)$ of the directions of particles immediately before their collision;

$\sigma_s(r)$ is the scattering cross-section coefficient;

$\sigma_a(r)$ is the absorption cross-section coefficient;

$\sigma(r)$ is the total extinction coefficient: $\sigma(r) = \sigma_s(r) + \sigma_a(r)$;

λ is the probability of the photon survival at a collision at the point r:

$$\lambda = \frac{\sigma_s(r)}{\sigma_s(r) + \sigma_a(r)};$$

$g(\omega' \cdot \omega)$ is the phase function (indicatrix);

f is the probability density of the free path length. In the standard case

$$f(l) = \sigma(r(l)) \exp(-\tau(l)),$$

where

$$\tau(l) = \int_0^l \sigma(r(t))dt$$

is the optical length of the free path;

$p(x)$ is the density of points x at the first collisions in the phase space.

As usual, we consider the linear functional

$$(F, \varphi) = \int_X F(x)\varphi(x)$$

of the collision density, which is a solution of the transport equation

$$F(r,\omega) = \int_\Omega d\omega' \, \frac{g(\omega' \cdot \omega)}{2\pi} \int_R dr' \frac{\exp(-\tau(r,r'))}{|r-r'|^2} \delta_s(\omega) F(r',\omega') + h(r,\omega). \tag{B.1}$$

Below, for simplicity we treat a medium with the extinction coefficient $\sigma(r) \equiv \sigma$ that does not depend on the spatial coordinate r. However, the results to be obtained remain true in the general case where the coefficient $\sigma(r)$ is not constant. In the considered case the density of the free path length has the form

$$f(l) = \sigma \exp(-\sigma l). \tag{B.2}$$

When $\sigma \to \infty$ this function weakly converges to the generalized δ-function. Define a small number $\varepsilon > 0$ and represent the density f as a linear combination of two functions normed to unity:

$$f = p f_\varepsilon + q\tilde{f}. \tag{B.3}$$

Here f_ε and $\tilde{f}$ are "δ-shaped" and "regular" components of the density f:

$$f_\varepsilon(t) = \begin{cases} \dfrac{f(t)}{p}, & 0 \le t \le \varepsilon, \\ 0, & \varepsilon < t < \infty, \end{cases}$$

$$\widetilde{f}(t) = \begin{cases} 0, & 0 \leq t \leq \varepsilon, \\ \dfrac{f(t)}{q}, & \varepsilon < t < \infty, \end{cases}$$

and the coefficients p and q are equal to

$$p = \int_0^{\varepsilon} f(t)dt, \qquad q = \int_{\varepsilon}^{\infty} f(t)dt.$$

The density $f_\varepsilon \to \delta$ as $\varepsilon \to 0$. Here δ is the Dirac function, the density is concentrated in the zero point:

$$\int_0^{\infty} \delta(t) f(t) dt = f(0).$$

At large values of the parameter σ we can use the approximate representation for the density f:

$$f(t) \simeq p\delta(t) + q\widetilde{f}(t) \tag{B.4}$$

We will discuss the error caused by this representation in Section 3.

So, we rewrite the transport equation (1) according to the representation (4) of the density of the free path length. Transform the integral term in the right-hand side of the equation. First, we rewrite the integral with respect to r. Denote

$$l = |r' - r|, \qquad s = \frac{r - r'}{|r - r'|}, \qquad |s| = 1, \qquad r' = r - l \cdot s$$

and change variables in this integral. We pass from the Cartesian coordinates (x, y, z) to the polar coordinates (l, s) $(0 \leq l < \infty,\ s \in \Omega)$. The Jacobian of the transformation $(x, y, z) \to (l, s)$ is equal to l^2. Now the integral can be rewritten in the following form:

$$\begin{aligned} &\int_R dr' \frac{\exp(-\tau(r, r'))}{|r - r'|^2} \delta_s(\omega) F(r', \omega') \\ &= \int_\Omega ds \int_0^\infty dl\, l^2 \frac{\exp(-\tau(l))}{l^2} \delta_s(\omega) F(r - l \cdot s, \omega') \\ &= \int_\Omega ds\, \delta_s(\omega) \int_0^\infty dl\ \exp(-\tau(l)) F(r - l \cdot s, \omega') \\ &= \int_0^\infty dl\ \exp(-\tau(l)) F(r - l \cdot \omega, \omega'). \end{aligned} \tag{B.5}$$

This is a one-dimensional integral over the ray $\{r' : r' = r - s \cdot \omega\}$. Now substitute (5) to the right-hand side of equation (1). We have

$$\int_\Omega d\omega' \frac{g(\omega' \cdot \omega)}{2\pi} \int_R dr' \frac{\exp(-\tau(r, r')}{|r - r'|^2} \delta_s(\omega) F(r', \omega') + h(r, \omega)$$

$$= \int_\Omega d\omega' \frac{g(\omega' \cdot \omega)}{2\pi} \int_0^\infty dl \exp(-\tau(l)) F(r - l \cdot \omega, \omega')$$

$$= \int_\Omega d\omega' \frac{g(\omega' \cdot \omega)}{2\pi} (p \cdot F(r, \omega') + q \cdot \int_0^\infty dl\, f(l) F(r - l \cdot \omega, \omega'))$$

$$= p \cdot \int_\Omega d\omega' \frac{g(\omega' \cdot \omega)}{2\pi} F(r, \omega') + q \cdot \int_\Omega d\omega' \frac{g(\omega' \cdot \omega)}{2\pi} \int_0^\infty dl\, f(l) F(r - l \cdot \omega, \omega')$$

$$F(r, \omega) = \int_\Omega d\omega' \frac{g(\omega' \cdot \omega)}{2\pi} \int_0^\infty dl \exp(-\tau(l)) F(r - l \cdot \omega, \omega'). \qquad \text{(B.6)}$$

Perform the substitution

$$f \to p\delta + q\widetilde{f}$$

and obtain successively:

$$\int_\Omega d\omega' \frac{g(\omega' \cdot \omega)}{2\pi} \int_R dr' \frac{\exp(-\tau(r, r'))}{|r - r'|^2} \delta_s(\omega) F(r', \omega') + h(r, \omega)$$

$$= \int_\Omega d\omega' \frac{g(\omega' \cdot \omega)}{2\pi} \int_0^\infty dl \exp(-\tau(l)) F(r - l \cdot \omega, \omega') + h(r, \omega)$$

$$= \int_\Omega d\omega' \frac{g(\omega' \cdot \omega)}{2\pi} (p \cdot F(r, \omega') + q \cdot \int_0^\infty dl\, f(l) F(r - l \cdot \omega, \omega'))$$

$$= p \int_\Omega d\omega' \frac{g(\omega' \cdot \omega)}{2\pi} F(r, \omega') + q \int_\Omega d\omega' \frac{g(\omega' \cdot \omega)}{2\pi} \int_0^\infty dl\, f(l) F(r - l \cdot \omega, \omega').$$

We have finished the necessary transformations in the right-hand side of the transport equation (1). The transformed equation takes the form

$$F(r,\omega)=p\int\limits_{\Omega} d\omega'\,\frac{g(\omega'\cdot\omega)}{2\pi}F(r,\omega')+q\int\limits_{\Omega} d\omega'\,\frac{g(\omega'\cdot\omega)}{2\pi}\int\limits_{0}^{\infty} dl\, f(l)F(r-l\cdot\omega,\omega'). \tag{B.1'}$$

Transform this equation into the standard form of the integral transport equation. Transpose the first integral from the right-hand side to the left-hand side and rewrite the equation in the compressed form:

$$\begin{aligned} F(r,\omega)-p\int\limits_{\Omega} d\omega'\,\frac{g(\omega'\cdot\omega)}{2\pi}F(r,\omega') \\ =q\int\limits_{\Omega} d\omega'\,\frac{g(\omega'\cdot\omega)}{2\pi}\int\limits_{0}^{\infty} dl\, f(l)F(r-l\cdot\omega,\omega')+h(r,\omega). \end{aligned} \tag{B.7}$$

For brevity denote by G the integral operator with kernel g. It acts on the functions defined on the sphere Ω:

$$G:\quad h(\omega)\to\int\limits_{\Omega} g(\omega\cdot\omega')h(\omega')d\omega'.$$

Using this notation rewrite equation (7):

$$F(r,\omega)-pGF(r,\omega)=qG\left[\int\limits_{0}^{\infty} dl\, f(l)F(r-l\cdot\omega,\omega')\right]+h(r,\omega) \tag{B.8}$$

or, expressing the left-hand side in operator form:

$$(E-pG)F(r,\omega)=qG\left[\int\limits_{0}^{\infty} dl\, f(l)F(r-l\cdot\omega,\omega')\right]+h(r,\omega). \tag{B.9}$$

Since $\|G\|<1$, we have $p<1$, and hence the inverse operator $(E-pG)^{-1}$ exists. So, equation (9) is equivalent to

$$F(r,\omega)=(E-pG)^{-1}qG\left[\int\limits_{0}^{\infty} dl\, f(l)F(r-l\cdot\omega,\omega')\right]+h(r,\omega). \tag{B.10}$$

It is known that the eigenfunctions of the operator G are the spherical functions $Y_l^m(\omega)$, $l = 0, 1, \ldots,$ $m = 0, \pm 1, \pm 2, \ldots,$ and each function $Y_l^m(\omega)$ corresponds to the eigenvalue g_l (Vladimirov, 1981). The operator $(E - pG)^{-1} qG$ has the same eigenfunctions as the operator G and the corresponding eigenvalues are equal to

$$\lambda_l = \frac{qg_l}{1 - pg_l}, \quad l = 0, 1, \ldots .$$

Besides, we will show that this operator is an integral one and its kernel is the function $\widetilde{g}(\omega, \omega') \equiv \widetilde{g}(\omega \cdot \omega')$, which, in fact, depends only on one scalar variable $\mu = (\omega \cdot \omega')$ and the function $\widetilde{g}(\mu)$ is a substochastic density (i. e. it is a nonnegative function and its integral does not exceed 1).

To prove this, one may study the functional series

$$\lambda_0 P_0(\mu) + \lambda_1 P_1(\mu) + \ldots ,$$

and, using the properties of the Legendre polynomials, show that this series converges when $-1 \le \mu \le 1$ and its sum is a substochastic density. But we can do it in a more simple way.

Represent the operator $(E - pG)^{-1}$ as a Neumann series:

$$(E - pG)^{-1} = E + pG + p^2 G^2 + \ldots .$$

Then

$$(E - pG)^{-1} \cdot qG = qG + pqG^2 + p^2 qG^3 \ldots .$$

Each operator G^n is an integral operator, and its kernel is equal to

$$g_n(\omega \cdot \omega') = \int\limits_{\Omega} \ldots \int\limits_{\Omega} g(\omega \cdot \omega_1) g(\omega_1 \cdot \omega_2) \cdot \ldots \cdot g(\omega_{n-1} \cdot \omega') d\omega_1 \ldots d\omega_{n-1}.$$

The kernel of the operator $(E - pG)^{-1} \cdot qG$ is equal to

$$q \cdot g_1 + pq \cdot g_2 + p^2 q g_3 + \ldots . \tag{B.11}$$

Further, if f and g are substochastic densities, that is,

$$f, g \ge 0, \qquad \int\limits_{\Omega} f(\omega \cdot \omega') d\omega' \le 1, \qquad \int\limits_{\Omega} g(\omega \cdot \omega') d\omega' \le 1,$$

then the function

$$h(\omega \cdot \omega') = \int\limits_{\Omega} f(\omega \cdot \omega'') g(\omega'' \cdot \omega') d\omega''$$

is a substochastic density too. We have $h \geq 0$ and

$$\int\limits_{\Omega}\left[\int\limits_{\Omega} f(\omega\cdot\omega'')g(\omega''\cdot\omega')d\omega''\right]d\omega' = \int\limits_{\Omega} f(\omega\cdot\omega'')\cdot\left[\int\limits_{\Omega} g(\omega''\cdot\omega')d\omega'\right]d\omega''$$

$$\leq \int\limits_{\Omega} f(\omega\cdot\omega'')\cdot 1\, d\omega'' \leq 1.$$

Hence, by induction we have proved that all the functions g_n are substochastic densities. Their linear combination (11) is also nonnegative function since $p, q > 0$, and its integral over the sphere Ω does not exceed

$$q + pq + p^2 q + \ldots = \frac{q}{1-p} = \frac{1-p}{1-p} = 1.$$

Therefore, the operator $(E - pG)^{-1}\cdot qG$ is an integral operator and its kernel $\widetilde{g}(\omega\cdot\omega')$ is substochastical density.

Hence, we can rewrite the transformed equation (9) as an integral equation of the second kind

$$F(r,\omega) = \int\limits_{\Omega}\frac{\widetilde{g}(\omega\cdot\omega')}{2\pi}\int\limits_{\varepsilon}^{\infty} dl\, \exp(-\tau(l))F(r - l\cdot\omega, \omega').$$

The obtained equation has the same form as the original one. We can regard it as a transport equation with the new phase function $\widetilde{g}$, the new density of the free path length $\widetilde{l}$ and the new source density function $\widetilde{h}$. Solving this modified transport equation, we should model photon trajectories according to these new parameters.

Note the differences in modelling the trajectories according to the old "natural" and the new "artificial" parameters.

The "artificial" free path length may vary within the limits $(0,\infty)$, and the "natural" one within the limits (ε,∞). Thus, the mean free path lengths become longer, the medium becomes more "permeable", its optical thickness decreases. Hence, the photons leave the transfer region in shorter time and their trajectories become shorter, that is, the average number of collisions per trajectory decreases. This results in shorter computation time.

Another difference is that the phase function becomes smoother: if there is a peak at $\mu = 1$, it diminishes. The probability of the photon survival at a collision increases:

$$\widetilde{g}_0 = g_0\frac{q}{1-pg_0} = g_0\frac{1-p}{1-pg_0} < g_0\frac{1-pg_0}{1-pg_0} = g_0.$$

This also results in decreasing waste of computer time and higher accuracy.

In the next section we give an illustrative explanation of the proposed method.

B.3. ILLUSTRATIVE STUDY OF THE METHOD

Let the contribution function $\varphi(x)$ be of the form

$$\varphi(r,\omega) = g(\omega \cdot \omega^*) \cdot C(r).$$

Such a function is often used in practical calculations.

First, we consider trajectory modelling according to the "natural" density. Let us choose a random trajectory $x_0, x_1, \ldots$ that is modelled according to that density. Then choose a small number $\varepsilon > 0$. If the free path length l becomes less than ε then we regard l as negligibly small. Write down a sequence of the spatial collision points $r_0, r_1, \ldots$. Let l_i be the distances between the neighbouring collision points:

$$l_i = |r_{i+1} - r_i|$$

Now, mark all the $l_i \leq \varepsilon$ the following way: $\bar{l}_i$. Since ε is very small, we can approximately assume that $\bar{l}_i = 0$ for all i. Thus, in the sequence $l_0, l_1, \ldots$ there appear subsequences that consist of zeros. Consider the spatial collision points r_i that are starting points of that subsequences. We will call them for brevity the "points of delay" because the particle stops moving after colliding at these points. Thus, the value of the sum of the contributions at these points is approximately equal to

$$C(r_i)(g(\omega_i \cdot \omega^*) + \ldots + g(\omega_{i+k} \cdot \omega^*)), \quad \bar{l}_i = \ldots = \bar{l}_{i+k} = 0.$$

Now we obtain the following way for trajectory modelling and functional calculation.

1. For a collision at a spatial point r_i we model a "stop" at the point r_i with probability $p = \exp(-\varepsilon)$ (that is, we model the free path length $l_i = 0$) and we model the free path length according to the density $\widetilde{f}$ with probability $q = 1 - p$. For "collisions" at the spatial "points of delay" r_i we calculate a value of the contribution function corresponding to the points (r_i, ω_{i+m}), $m = 0, \ldots, k$ of the phase space that have the same spatial coordinates r_i and different direction coordinates ω_{i+m}. Thus, in modelling trajectories we can economize on modelling the free path lengths l_{i+m}, $m = 0, \ldots, k$.

2. We will go further than that. We will find the mathematical expectation of the sum

$$g(\omega_i \cdot \omega^*) + \ldots + g(\omega_{i+k} \cdot \omega^*). \tag{B.12}$$

If a successive collision takes place at a point r_i, then with probability p the particle will suffer a "delay" at the point r_i and with probability q it will go as long as $l_i > \varepsilon$. In the first case the particle will change the "flight direction" $\omega_i \to \omega_{i+1}$ (though it will stay at the same spatial point r_i), and with probability p it will suffer a "delay" at the point r_i, and with probability q it will go as long as $l_i > \varepsilon$, and so on. Thus, mathematical expectaion of the sum (12) equals

$$g(\omega_i \cdot \omega^*) + p \int_\Omega g(\omega_i \cdot \omega_{i+1}) g(\omega_{i+1} \cdot \omega^*)$$
$$+ p^2 \int_\Omega g(\omega_i \cdot \omega_{i+1})\, g(\omega_{i+1} \cdot \omega_{i+2})\, g(\omega_{i+2} \cdot \omega^*) + \ldots. \tag{B.13}$$

Each term in this sum depends on a scalar $\mu = \omega_i \cdot \omega^*$, not on two vectors ω_i, ω^*. Therefore, the whole sum depends on the scalar μ. We can calculate this sum finding the coefficients k_n of its decomposition into the series over the Legendre polynomials.

The coefficients k_n may be expressed through the relevant coefficients g_n of the phase function $g(\mu)$. Taking into account that the calculation result should be multiplied by q, this expression takes the form

$$k_n = g_n \frac{q}{1 - p g_n}.$$

Finally, we obtain the following way for the calculation. In modelling the trajectory we use the free path density $\tilde{f}$ and the phase function $\tilde{g}$ instead of the old f and g, respectively.

The advantage is that the sum (12) is calculated analytically. In the corresponding Monte Carlo calculation we are to model the relevant free path lengths and the values of the phase function. It is clear that the advantage increases as the extinction coefficient σ grows, and correspondingly grows the probability p of the "delay".

Besides, the "artificial" phase function $\tilde{g}$ is smoother than the original one since the Fourier coefficients in a decomposition into the series over the Legendre polynomials for $\tilde{g}$ are less then the corresponding coefficients for g. This also diminishes the computation error.

However, it is not recommended to use the proposed method in case of highly peaked phase function (see Section 5).

B.4. ESTIMATION OF ACCURACY

The error of the calculation by the proposed method arises from the distortion in modelling the trajectories according to the new artificial parameters. As the optical thickness of the medium increases, the number of collisions per trajectory along with the number of the points of delay increases too. In this respect a question arises: whether the error of the calculation will accumulate or not.

Let us try to estimate the error. Consider equation (1′)

$$F(r,\omega) = \int_{\Omega} d\omega' \, \frac{g(\omega' \cdot \omega)}{2\pi} \int_0^{\infty} dl \, \exp(-\tau(l)) F(r - l \cdot \omega, \omega') + h(r,\omega) \quad \text{(B.14)}$$

and transform it.

First we use the relations

$$F(r,\omega) = p \cdot F(r,\omega) + q \cdot F(r,\omega)$$

and

$$\int_0^{\infty} dl \, \exp(-\tau(l)) F(r - l \cdot \omega, \omega') = p \cdot \int_0^{\infty} \tilde{p}(t) F(t)\, dt + q \cdot \int_0^{\infty} \tilde{q}(t) F(t)\, dt.$$

Then we obtain

$$p \cdot F(r,\omega) + q \cdot F(r,\omega)$$
$$= \int_{\Omega} d\omega' \, \frac{g(\omega' \cdot \omega)}{2\pi} \left[p \cdot \int_0^{\infty} \tilde{p}(t) F(t)\, dt + q \cdot \int_0^{\infty} \tilde{q}(t) F(t)\, dt \right] + h(r,\omega).$$

Substitute the following expression for $F(r,\omega)$:

$$F(r,\omega) = \int_{\Omega} d\omega' \, \frac{g(\omega' \cdot \omega)}{2\pi} F(r,\omega).$$

We have

$$p \cdot \int_{\Omega} d\omega' \, \frac{g(\omega' \cdot \omega)}{2\pi} F(r,\omega) + q \cdot F(r,\omega)$$
$$= \int_{\Omega} d\omega' \, \frac{g(\omega' \cdot \omega)}{2\pi} \left[p \cdot \int_0^{\infty} \tilde{p}(t) F(t)\, dt + q \cdot \int_0^{\infty} \tilde{q}(t) F(t)\, dt \right] + h(r,\omega)$$

then

$$p \cdot \int_{\Omega} d\omega' \frac{g(\omega' \cdot \omega)}{2\pi} F(r,\omega) + q \cdot F(r,\omega)$$

$$= q \cdot \int_{\Omega} d\omega' \frac{g(\omega' \cdot \omega)}{2\pi} \int_0^{\infty} \widetilde{q}(t) F(t) dt$$

$$+ p \cdot \int_{\Omega} d\omega' \frac{g(\omega' \cdot \omega)}{2\pi} \int_0^{\infty} \widetilde{p}(t) F(t) dt + h(r,\omega).$$

Transpose the first term in the left-hand side of the equation to the right, rearrange the terms and divide both sides of the equation by q. Then we obtain

$$q \cdot F(r,\omega) = q \cdot \int_{\Omega} d\omega' \frac{g(\omega' \cdot \omega)}{2\pi} \int_0^{\infty} \widetilde{q}(t) F(t)\, dt$$

$$+ p \cdot \int_{\Omega} d\omega' \frac{g(\omega' \cdot \omega)}{2\pi} \cdot \left[\int_0^{\infty} \widetilde{p}(t) F(t)\, dt - F(r,\omega) \right] + h(r,\omega).$$

Finally, equation (1′) can be rewritten in the following equivalent form:

$$F(r,\omega) = \int_{\Omega} d\omega' \frac{g(\omega' \cdot \omega)}{2\pi} \int_0^{\infty} \widetilde{q}(t) F(t)\, dt + \frac{1}{q} h(r,\omega)$$

$$+ \frac{p}{q} \cdot \int_{\Omega} d\omega' \frac{g(\omega' \cdot \omega)}{2\pi} \cdot \left[\int_0^{\infty} \widetilde{p}(t) F(t)\, dt - F(r,\omega) \right]. \tag{B.15}$$

The proposed method is based on the assumption that the difference

$$\int_0^{\infty} \widetilde{p}(t) F(t)\, dt - F(r,\omega)$$

is approximately equal to zero. Starting from this assumption we obtain the new equation for the approximate collision density $\widetilde{F}$:

$$\widetilde{F}(r,\omega) = \int_{\Omega} d\omega' \frac{g(\omega' \cdot \omega)}{2\pi} \int_0^{\infty} \widetilde{q}(t) \widetilde{F}(t)\, dt + \frac{1}{q} h(r,\omega). \tag{B.16}$$

Thus, we have obtained two equations for the functions F and $\widetilde{F}$, correspondingly. Their difference has the form

$$F(r,\omega) - \widetilde{F}(r,\omega) = \int_{\Omega} d\omega' \, \frac{g(\omega' \cdot \omega)}{2\pi} \int_{0}^{\infty} \widetilde{q}(t)(F(r,\omega) - \widetilde{F}(r,\omega)) \, dt$$
$$+ \frac{p}{q} \cdot \int_{\Omega} d\omega' \, \frac{g(\omega' \cdot \omega)}{2\pi} \cdot \left[\int_{0}^{\infty} \widetilde{p}(t) F(t) \, dt - F(r,\omega) \right] . \tag{B.17}$$

We may regard it as an integral equation for the function $F - \widetilde{F}$ with the source density

$$\frac{p}{q} \cdot \int_{\Omega} d\omega' \, \frac{g(\omega' \cdot \omega)}{2\pi} \cdot \left[\int_{0}^{\infty} \widetilde{p}(t) F(t) \, dt - F(r,\omega) \right] .$$

We only need to show that if the source density function is small, then the solution of equation (17) is small. In fact, we are interested not in the collision density, which increases as the optical thickness increases. Instead, we are interested in the density of the flux.

But it is well known that the density of the flux depends on the source density in continuous way. Thus we see that the accuracy of the proposed method does not increase as the optical thickness increases.

B.5. LIMITATION ON THE USE OF THE METHOD

The applicability of the proposed method has natural limitations.

1. We cannot use the new parameters for the trajectory modelling if the particle is close to the medium boundary. In this case the probability that the particle leaves the medium is high. However, modelling the free path length according to the new density $\widetilde{f}$ artificially prolongs its lifetime, which results in overrating the estimation.

To avoid this effect we should use the old "natural" parameters in modelling trajectories when the collision point is in a δ-vicinity of the boundary surface and we should use the new "artificial" parameters outside this vicinity. The values ε and δ should be chosen on the basis of practical considerations. Proper theoretical estimation is unavailable here.

2. Apparently, the proposed method will give a poor result in the case of highly peaked phase function. In this case, the real particle will go far from

a point of delay with large probability, while the artificial particle would stay at the point after a number of collisions.

B.6. NUMERICAL EXPERIMENT

To test the efficiency of the proposed method, a model numerical experiment has been carried out. A radiation transfer process without absorption has been considered in a plane layer of unit geometrical thickness. The scattering coefficient inside the layer is equal to 10.

The matter outside the layer absorbs all the outgoing radiation. The phase function is assumed isotropic. A monodirectional radiation source is mounted on the boundary plane of the layer. The problem was to calculate the outgoing flux of the radiation which leaves the layer through the other boundary plane. The direction of the source is $(0, 0, 1)$, the observation direction is $(\sqrt{2}/2, 0, \sqrt{2}/2)$.

In calculation we have taken $\varepsilon = 0.1$. Two thin δ-layers are adjacent to the boundary planes. In calculation we have taken $\delta = 0.1$.

It is clear that in the case of isotropic phase function $g(\mu) \equiv 1/2$ the coefficients in the decomposition of g and $\widetilde{g}$ are equal, and hence $g(\mu) \equiv 1/2 \equiv \widetilde{g}$. Therefore, in this example the "artificial" phase function coincides with the "natural" one. The error of the computation by the modified method was the same as the error of the standard method. But the computation time for the standard method is twice as much as the time for the modified method for the same number of trajectories.

Appendix C.

A new method for identification of the phase function

The method for identification of the phase function considered in the preceding sections worked for τ up to 0.8. But we failed to obtain satisfactory results with proper accuracy at larger values of τ. Therefore, a new method has been developed for solving this problem.

In brief, the approach is as follows. The proposed method is based on a transformation of the transport equation. The method is iterative. At each iteration one should solve a system of linear equations for the components of a vector that defines a successive approximation for the desired phase function. Coefficients of that system are linear functionals of the solution of the transport equation. They can be found by the Monte Carlo method. We present a brief description for a method of calculating these coefficients. Finally, some numerical results are presented.

C.1. REMARKS CONCERNING NOTATION

We use a piecewise approximation for the phase function $g(\mu)$ with given values

$$g(\mu_i) = g_i \geq 0, \quad i = 1, \ldots, n$$

at some points (nodes) μ_i in the interval $[-1, +1]$:

$$1 = \mu_1 > \mu_2 > \ldots > \mu_n = -1.$$

The phase function is interpolated linearly between the points. We may present the piecewise function $g(\mu)$ as a linear combination of the basic functions $h_k(\mu)$:

$$g(\mu) = g_1 h_1(\mu) + \ldots + g_n h_n(\mu), \tag{C.1}$$

where

$$h_1(\mu) = \begin{cases} \dfrac{\mu - \mu_2}{1 - \mu_2}, & \mu_2 \le \mu \le 1, \\ 0, & \text{otherwise} \end{cases}$$

$$h_n(\mu) = \begin{cases} \dfrac{\mu_{n-1} - \mu}{\mu_{n-1} + 1}, & -1 \le \mu \le \mu_{n-1}, \\ 0, & \text{otherwise} \end{cases}$$

$$h_i(\mu) = \begin{cases} \dfrac{\mu - \mu_{i+1}}{\mu_i - \mu_{i+1}}, & \mu_{i+1} \le \mu \le \mu_i, \\ \dfrac{\mu_{i-1} - \mu}{\mu_{i-1} - \mu_i}, & \mu_i \le \mu \le \mu_{i-1} \end{cases} \qquad i = 2, \ldots, n-1$$

Thus, on a fixed set of the nodes $\mu_1, \mu_2, \ldots, \mu_n$ the phase function $g(\mu)$ can be specified by a vector

$$g = (g_1, g_2, \ldots, g_n).$$

In turn, an arbitrary vector $(g_1, g_2, \ldots, g_n)$ corresponds to a function $g(\mu)$. In calculations by the Monte Carlo method it is more convenient to use the decomposition as follows

$$g(\mu) = (g_1 \|h_1\|)\overline{h}_1(\mu) + \ldots + (g_n \|h_n\|)\overline{h}_n(\mu), \tag{C.2}$$

where $\overline{h}_k$ are the normed basic functions:

$$\overline{h}_i(\mu) = h_i(\mu)/\|h_i\|, \qquad \|h_i\| = \int_{-1}^{+1} h(\mu)d\mu.$$

The functions $\overline{h}_k$ are the probability density functions. It is easy to see that

$$\|h_1\| = \frac{1 - \mu_2}{2}, \qquad \|h_n\| = \frac{\mu_{n-1} + 1}{2},$$

$$\|h_i\| = \frac{\mu_{i-1} - \mu_{i+1}}{2}, \qquad i = 2, \ldots, n-1.$$

Below we use the following norms for vectors, matrices and functions, respectively:

$$\|x\| = |x_1| + |x_2| + \ldots + |x_n|,$$

$$\|(a_{i,j})\| = \max_j \sum_{i=1}^{n} |a_{i,j}|qq\|f(x)\| \equiv \sup_x |f(x)|.$$

C.2. PROBLEM STATEMENT

A medium that scatters and absorbs the radiation is illuminated by a source having distribution density $p(r, \omega)$. Radiation detectors are located at the spatial points $r_1, \ldots, r_n$. The measured values of the scattered intensity coming into the point r_i in the direction ω_i are $F_i = F(r_i, \omega_i)$. The problem is to find the phase function at the given values F_i, $i = 1, \ldots, n$.

C.3. ITERATIVE METHOD

In further considerations we will use the transport equation

$$F(r, \omega) = \int_{\Omega} d\omega' \frac{g(\omega' \cdot \omega)}{2\pi} \int_{R} dr' \frac{\sigma(r) \exp(-\tau(r, r'))}{|r - r'|^2} \delta_\omega(s(r', r)) F(r', \omega') + p(r, \omega), \tag{C.3}$$

which may be rewritten in the following form:

$$F(r, \omega) = \int_{-1}^{+1} d\mu' g(\mu') \cdot \frac{1}{2\pi} \int_{0}^{2\pi} d\varphi' \int_{0}^{\infty} dt\, \sigma(r) \exp(-\tau(t))\; F(r - t \cdot \omega, \omega') + p(r, \omega). \tag{C.4}$$

This is the equation for the unknown collision density F with known other parameters, including the phase function. However, this equation may be regarded as an equation for the unknown phase function $g(\mu)$ with known values $F(r, \omega)$ in the left-hand side of this equation. But if we try to find the phase function from this equation, we will face the following obstacle. To find n unknown components $g_1, \ldots, g_n$, we need the same number of known values $F(r_1, \omega_1), \ldots, F(r_n, \omega_n)$, (these are the measured values of the radiation intensity at certain points of the phase space). But the transport equation involves infinite number of unknown values $F(r, \omega)$. Therefore, we fail to obtain a closed system of equations for the needed phase function $g(\mu)$, even when we need to know not the whole function but only n values $(g_1, g_2, \dots, g_n)$.

We will use another way of solving this equation. Now we will obtain not a direct but an iterative process for estimating the needed function g.

First, we define a function $F_x(\mu)$, which is the integral of the collision density $F(x)$:

$$F_x(\mu') = \frac{1}{2\pi} \int_0^{2\pi} d\varphi' \int_0^{\infty} dt\, \sigma(r) \exp(-\tau(t))\, F(r - t \cdot \omega, \omega'). \tag{C.5}$$

Now the transport equation can be rewritten in the form

$$F(r, \omega) = \int_{-1}^{+1} d\mu' g(\mu') \cdot F_x(\mu') + p(r, \omega). \tag{C.6}$$

Here we substitute the expression (1) for the function $g(\mu')$ and obtain

$$F(r, \omega) = \sum_{j=1}^{n} g_j \int_{-1}^{+1} d\mu' h_j(\mu') F_x(\mu') + p(r, \omega). \tag{C.7}$$

Now we fix the spatial points $r^1, r^2, \dots$, where the radiation detectors are located, and the relevant observation directions $\omega^1, \omega^2, \dots$. There is a correspondence between the detectors $x^i = (r^i, \omega^i)$ and intensities of the observed scattered radiation $F(r^i, \omega^i)$. For each $x^i = (r^i, \omega^i)$ the relation (7) may be rewritten as follows:

$$\sum_{j=1}^{n} a_{ij} g_j = b_i, \quad i = 1, \dots, n, \tag{C.8}$$

where

$$a_{ij} = \int_{-1}^{+1} d\mu'\ h_j(\mu') F_{x^i}(\mu'), \qquad b_i = F(r^i, \omega^i) - p(r^i, \omega^i). \tag{C.9}$$

Finally, we obtain a system of linear equations (8) for the components $g_1, \dots, g_n$.

However, in this system only the right-hand side (b_i) is known, but the coefficients (a_{ij}) of the system are still unknown. They depend on the values (g_i) that are to be found:

$$a_{ij} = a_{ij}(g).$$

Therefore, the system of equations (8) can be rewritten as follows:

$$\sum_{j=1}^{n} a_{ij}(g) g_j = b_i, \quad i = 1, \ldots, n. \tag{C.10}$$

We proceed the following way. Let $g^*(\mu)$ be the desired phase function. Then $g^* \sim (g_1^*, \ldots, g_n^*)$ and, correspondingly, $a_{ij}^* = a_{ij}(g^*)$.

To find the vector $(g_1^*, \ldots, g_n^*)$, we build up an iterative process. Let $(g_1^0, \ldots, g_n^0)$ be some initial approximation to the desired vector $(g_1^*, \ldots, g_n^*)$. Then in the system of linear equations (10) replace the unknown values a_{ij}^* with their approximations $a_{ij}^0 = a_{ij}(g^0)$.

Solving the system of linear equations

$$\sum_{j=1}^{n} a_{ij}^0 g_j^1 = b_i, \quad i = 1, \ldots, n$$

we obtain the following approximation $(g_1^1, \ldots, g_n^1)$ for the desired vector. Then we calculate the following approximation $a_{ij}^1 = a_{ij}(g^1)$ for the coefficients a_{ij} of the system, and solving the system

$$\sum_{j=1}^{n} a_{ij}^1 g_j^2 = b_i, \quad i = 1, \ldots, n$$

we find the next approximation to the desired vector $(g_1^2, \ldots, g_n^2)$, and so on.

In other words, if the k-th iteration g^k is known then the following iteration g^{k+1} can be found from the system of linear equations that defines the iterative process

$$\sum_{j=1}^{n} a_{ij}^k g_j^{k+1} = b_i, \quad i = 1, \ldots, n. \tag{C.11}$$

If certain conditions are satisfied, then $g^k \to g^*$ at $k \to \infty$. We will discuss it below.

The most effort-consuming part of the described method consists in computing coefficients a_{ij}^k of the matrix of the system of linear equations. To calculate them, we have to solve the transport equation (3). On calculating the coefficients a_{ij} see the next section.

C.4. CALCULATION OF THE MATRIX COEFFICIENTS

Our consideration is focused on the new method for finding the phase function. Therefore, we will not go far into the details of the calculating method and discuss its advantages and disadvantages. We only consider one Monte Carlo way for calculating the coefficients a_{ij}.

According to the conventional scheme of the Monte Carlo method, we represent a_{ij} as a linear functional of the solution of the transport equation

$$(F, \varphi) = \int\limits_X F(x)\varphi(x)$$

(we will find the proper function φ below). Then we model Markov chains $x_1, x_2, \dots, x_n$ of successive collision points (x_n is the point of the trajectory termination). At each collision point x_m we calculate $\varphi(x_m)$ and the "weight" of the particle Q_m:

$$Q_m = Q_{m-1} \cdot \frac{k(x_{m-1} \to x_m)}{\overline{k}(x_{m-1} \to x_m)},$$

where k is the kernel of the transport equation (4) (a "natural" density for modelling a successive collision point), and $\overline{k}$ is an "artificial" density. The sum

$$\xi = Q_1\varphi(x_1) + Q_2\varphi(x_2) + \ldots + Q_n\varphi(x_n)$$

gives a statistical estimate of the contribution of the whole trajectory x_1, $x_2, \dots, x_n$. The desired linear functional is equal to mathematical expectation of the random value ξ over all trajectories $x_1, x_2, \dots, x_n$:

$$(F, \varphi) = E\xi.$$

Thus, to perform calculation, we need to know probability densities for modelling trajectories (i. e. densities for the free path lengths and path directions), the contribution function, and the factors Q_i.

Rewrite the expression for the coefficients a_{ij} in more details:

$$\begin{aligned} a_{ij} &= \int\limits_{-1}^{+1} d\mu'\, h_j(\mu') F_{\omega^i}(\mu') \\ &= \int\limits_{-1}^{+1} d\mu'\, h_j(\mu') \frac{1}{2\pi} \int\limits_0^{2\pi} d\varphi' \int\limits_0^{\infty} dt\, \sigma(r) \exp(-\tau(t))\, F(r - t \cdot \omega^i, \omega') \end{aligned}$$

$$= \int_{\Omega} d\omega' \, \frac{h_j(\omega \cdot \omega')}{2\pi} \int_0^{\infty} dt \, \sigma(r) \exp(-\tau(t)) \, F(r - t \cdot \omega^i, \omega')$$

$$= \|h_j\| \int_{\Omega} d\omega' \, \frac{\overline{h}_j(\omega \cdot \omega')}{2\pi} \int_0^{\infty} dt \; \sigma(r) \exp(-\tau(t)) \; F(r - t \cdot \omega^i, \omega') \tag{C.12}$$

It is clear that the coefficients a_{ij} are the linear functionals of the solution of the transport equation. It can be seen from the expression (11) that we may calculate them by the Monte Carlo method, modelling trajectories for an equation adjoint to the transport equation (3).

Let us consider probability densities for modelling these trajectories.

The particle starts from the spatial point r where the radiation detector is located. The starting direction is $\omega_0^i \equiv \omega^i$. The initial particle "weight" equals $\|h_j\|$. The probability density for the first free path length coincides with one that we use in the "natural" modelling:

$$f(l) = \sigma(r(l)) \exp(-\tau(l)).$$

We can take $\overline{h}_j(\omega_0^i \cdot \omega_1^i)$ as the density for modelling the path direction ω_1^i. In this case we have to model different trajectories for every coefficient a_{ij} (i is fixed, j varies from 1 to n). However, there is a way to reduce the amount of computation.

Let us model the path direction ω_1^i according to the usual density $g(\omega_0^i \cdot \omega_1^i)$ that we use in modelling particle trajectory. But the particle "weight" Q should be multiplied by a special factor:

$$Q \to Q \cdot \frac{\overline{h}_j(\omega_0^i \cdot \omega_1^i)}{g(\omega_0^i \cdot \omega_1^i)} = Q \cdot \frac{\overline{h}_j(\mu)}{g(\mu)}.$$

Thus, modelling one trajectory we are able to calculate statistical estimates for all the coefficients a_{ij} (i is fixed, j varies from 1 to n) at the same time. The following free path lengths and path directions should be modelled the same way as in the case of the standard particle trajectory.

So, to calculate the coefficients a_{ij} one may model the standard "adjoint trajectories" that start at the point of the detector, but after modelling the first scattering act the particle "weight" should be multiplied by a specific factor.

C.5. CONVERGENCE OF THE ITERATIONS

In this subsection we will obtain sufficient conditions for the proposed iterative method to converge.

Consider the transition from an iteration g_k to g^{k+1} and find a sufficient condition which gives a guarantee that the inequality

$$\|g^{k+1} - g^*\| \leq C\|g^k - g^*\|$$

holds, where the constant $C < 1$. If this inequality holds, then the considered iterative process will converge.

Denote by A_g the square matrix having the coefficients a_{ij} taken from the system of equations (11). We denote its right-hand side by b. Vectors g^k, g^{k+1} and g^* satisfy the equations

$$A_{g^k} g^{k+1} = b, \qquad A_{g^*} g^* = b.$$

Subtract the second equation from the first one. Then we obtain successively:

$$\begin{aligned} A_{g^k} g^{k+1} - A_{g^*} g^* &= (A_{g^k} g^{k+1} - A_{g^k} g^*) + (A_{g^k} g^* - A_{g^*} g^*) \\ &= A_{g^k}(g^{k+1} - g^*) + (A_{g^k} - A_{g^*})g^* = 0. \end{aligned}$$

Hence,

$$A_{g^k}(g^{k+1} - g^*) = (A_{g^*} - A_{g^k})g^*$$

and then

$$g^{k+1} - g^* = A_{g^k}^{-1}(A_{g^*} - A_{g^k})g^*.$$

This results in the inequalities for the norms (on the used norms see Section 1):

$$\|g^{k+1} - g^*\| = \|A_{g^k}^{-1}(A_{g^*} - A_{g^k})g^*\| \leq \|A_{g^k}^{-1}\| \|A_{g^*} - A_{g^k}\| \|g^*\|. \qquad \text{(C.13)}$$

All the values $\|A_{g^k}^{-1}\|$ are uniformly bounded by a constant C_A if the functions g^k are contained in a bounded vicinity of g^*.

Now consider two arbitrary functions g^1 and g^2. For brevity, let us take the following notation. We denote by K_1, K_2 the integral operators of the transport equation containing the functions g^k $(k = 1, 2)$. Then we mark by upper k all the parameters and functions that correspond to the function g^k:

$$a_{ij}^k = a_{ij}(g^k), \qquad F^k = F(g^k), \qquad F_i^k = F_{x^i}(g^k), \qquad \text{etc.} \qquad \text{(C.14)}$$

Now estimate the norm of the difference

$$\|A_{g^1} - A_{g^2}\| \equiv \|A^1 - A^2\|.$$

We have

$$\|A^1 - A^2\| \equiv \max_{1\le i\le n} \sum_{j=1}^{n} |a^1_{ij} - a^2_{ij}| \le n \cdot \max_{1\le i,j\le n} |a^1_{ij} - a^2_{ij}|. \tag{C.15}$$

Let

$$C_h = \max_j \|h_j\|. \tag{C.16}$$

Using the definition (13) for a_{ij}, estimate the value $|a^1_{ij} - a^2_{ij}|$, $i, j = 1, \dots, n$:

$$|a^1_{ij} - a^2_{ij}| = \left| \int_{-1}^{+1} d\mu \; h_j(\mu)(F^1_i(\mu) - F^2_i(\mu)) \right| \le C_h \cdot \left| \int_{-1}^{+1} d\mu \; (F^1_i(\mu) - F^2_i(\mu)) \right|. \tag{C.17}$$

Now we use the expression (5) that defines $F_x(\mu)$, and the denotation (16):

$$\begin{aligned}
&C_h \cdot \left| \int_{-1}^{+1} d\mu \, (F^1_i(\mu) - F^2_i(\mu)) \right| \\
&= C_h \cdot \left| \int_{-1}^{+1} d\mu \frac{1}{2\pi} \int_0^{2\pi} d\varphi \int_{-\infty}^{0} dt \, \sigma(r_i) \exp(-\tau(t)) \right. \\
&\qquad \left. \times (F^1(r_i - t\cdot\omega^i, \omega') - F^2(r_i - t\cdot\omega^i, \omega')) \right| \\
&\le 2C_h \cdot \sup_{t,\omega} |F^1(r_i - t\cdot\omega^i, \omega) - F^2(r_i - t\cdot\omega^i, \omega)| \\
&\le 2C_h \cdot \sup_{r,\omega} |F^1(r, \omega) - F^2(r, \omega)|.
\end{aligned} \tag{C.18}$$

Let us obtain an estimation for

$$\|F^1 - F^2\| \equiv \sup_{r,\omega} |F^1(r, \omega) - F^2(r, \omega)|.$$

Here

$$F^i = (E - K_i)^{-1}, \tag{C.19}$$

$$\|F^1 - F^2\| = \|(E - K_1)^{-1} - (E - K_2)^{-1}\|.$$

Use the representation of the operator $(E-K)^{-1}$ as a Neumann series:

$$(E-K)^{-1} = E + K + K^2 + \dots$$

We have

$$(E-K_1)^{-1} - (E-K_2)^{-1} = (K_1 - K_2) + (K_1^2 - K_2^2) + \dots$$

Hence,

$$\|(E-K_1)^{-1} - (E-K_2)^{-1}\| \le \|K_1 - K_2\| + \|K_1^2 - K_2^2\| + \dots. \tag{C.20}$$

Now we prove the following

Theorem 2. *Let $q = \max(\|K_1\|, \|K_2\|)$. Then the following inequality holds*

$$\|K_1^m - K_2^m\| \le 2\|g_1 - g_2\| \cdot mq^{m-1}. \tag{C.21}$$

Proof. The proof will be performed by induction over m. First, write down two inequalities for the norm of the integral operator of the transport equation $\|K\|$.

1. If radiation is transferred through a bounded region, then $\|K\| < 1$. Hence, $q < 1$.

2. Let us obtain the second inequality, which is suitable when the value $\|g\|$ is small. We use the following simple inequalities.

For an arbitrary function $f(\omega)$ we have:

$$\left|\int_\Omega d\omega' \, \frac{g(\omega\cdot\omega')}{2\pi} f(\omega')\right| = \left|\frac{1}{2\pi}\int_0^{2\pi} d\varphi' \int_{-1}^{+1} d\mu' g(\mu') f(\omega)\right| \le 2\|g\|\|f\|.$$

Then, for an arbitrary function $f(t)$

$$\left|\int_0^\infty dt\, \sigma(0)\exp(-\tau(t))\, f(t)\right| \le \left|\int_0^\infty d\exp(-\tau(t))\right| \cdot \|f(t)\| \le \|f\|.$$

It follows from these two inequalities that

$$\left|\int_\Omega d\omega' \, \frac{g(\omega\cdot\omega')}{2\pi} \int_0^\infty dt\, \sigma(r)\exp(-\tau(t))\, F(r - t\cdot\omega, \omega')\right|$$

$$\le 2\|g\| \left\|\int_0^\infty dt\, \sigma(r)\exp(-\tau(t))\, F(r - t\cdot\omega, \omega')\right\| \le 2\|g\|\|F\|.$$

Hence

$$\|K\| \leq 2\|g\|. \tag{C.22}$$

Returning to the proof of the inequality (21) we write an estimation for $\|K_1 - K_2\|$:

$$K_1 f(r,\omega) - K_2 f(r,\omega) = \int_\Omega d\omega' \, \frac{g_1(\omega\cdot\omega') - g_2(\omega\cdot\omega')}{2\pi} \times \int_0^\infty dt \, \sigma(r)\exp(-\tau(t)) \, f(r - t\cdot\omega, \omega').$$

According to (22), we have

$$\|K_1 - K_2\| \leq 2\|g_1 - g_2\|.$$

Hence, the inequality (21) is now proved for $m = 1$. Let it be proved for $m = n$. Then for $m = n+1$ we have:

$$\begin{aligned} K_1^{n+1} - K_2^{n+1} &= K_1^{n+1} - K_1 \cdot K_2^n + K_1 \cdot K_2^n - K_2^{n+1} \\ &= K_1 \cdot (K_1^n - K_2^n) + (K_1 - K_2)\cdot K_2^n. \end{aligned}$$

Hence,

$$\begin{aligned} \|K_1^{n+1} - K_2^{n+1}\| &\leq \|K_1\| \cdot \|K_1^n - K_2^n\| + \|K_1 - K_2\| \cdot \|K_2\|^n \\ &\leq q \cdot 2nq^{n-1}\|g_1 - g_2\| + 2\|g_1 - g_2\| \cdot q^n \\ &= 2(n+1)q^n\|g_1 - g_2\|. \end{aligned}$$

Thus, the inequality (21) is proved. □

We return to the inequality for the norm $\|F^1 - F^2\|$. In accordance with (19) and (20), we have:

$$\begin{aligned} \|F^1 - F^2\| &\leq \|(E - K_1)^{-1} - (E - K_2)^{-1}\| \\ &\leq \|K_1 - K_2\| + \|K_1^2 - K_2^2\| + \ldots \\ &\leq 2\|g_1 - g_2\| + 4q\|g_1 - g_2\| + \ldots + 2mq^{m-1}\|g_1 - g_2\| + \ldots \\ &= 2\|g_1 - g_2\| \cdot (1 + 2q + \ldots + mq^{m-1} + \ldots) \\ &= 2\|g_1 - g_2\| \cdot \frac{d}{dq}(1 + q^2 + \ldots + q^n + \ldots) \\ &= 2\|g_1 - g_2\| \cdot \frac{1}{(1-q)^2}. \end{aligned} \tag{C.23}$$

It follows from (15), (21), (22) and (23) that

$$\|A^1 - A^2\| \leq n \cdot \max_{1\leq i,j\leq n} |a^1_{ij} - a^2_{ij}| \leq \frac{4nC_h}{1-q^2}\|g_1 - g_2\|, \tag{C.24}$$

and finally, from (13) and (24) we have

$$\|g^{k+1} - g^*\| \leq \frac{4nC_hC_A\|g^*\|}{1-q^2} \cdot \|g^k - g^*\|.$$

So, if the condition

$$\frac{4nC_hC_A\|g^*\|}{(1-q)^2} < 1$$

holds, then the iterative process (11) converges.

C.6. RESULTS OF NUMERICAL EXPERIMENT

A numerical experiment has been carried out to examine the efficiency of the proposed method. We compared the efficiency of the new method and the old one (see Sections 1 and 2).

We solved the following model problem. A plane homogeneous scattering layer is considered. Its upper boundary is illuminated by a flux of monodirectional solar radiation.

The observation scheme has been taken the same as in Section 1. The radiation detector is located on the earth's surface at a point D. It measures intensities of scattered solar radiation in several directions that make up equal zenith angles $\theta = \theta_\odot$ ($\theta_\odot$ is the solar zenith angle) and different azimuth angles φ (φ is the angle between the vertical plane of observation direction and the plane of the solar vertical, $0° < \varphi < 180°$).

The angle of single scattering ranges from zero to $2\theta < 180°$ and the cosines of the once-scattering angles coincide with the nodes μ_i of the phase function.

The "measured" intensity values have been computed in advance by the Monte Carlo method. To eliminate the influence of the "measurement" error on the quality of the identification, the computation has been carried out with high accuracy, so that the statistical error of computation did not exceed 1%.

Here we present results of numerical experiments in the identification of the phase function in a layer having optical thickness $\tau = 0.8$ and $\tau = 1.5$.

In the experiment we have considered a scattering medium without absorption. A model function has been taken from Radiation Commission IAMAP, WMO.WCP-112 (1986). In both cases $\theta_{\odot} = 80°$. The difference between the number of the nodes and the number of observation directions is equal to 1. We use the norming equation

$$\int_{-1}^{+1} g(\mu) d\mu = 1.$$

as an additional equation. Note that in the proposed method the points and the observation directions could be chosen arbitrarily. Unlike the old method (Section 1), the difference between the number of unknown parameters and the measured values is connected only with the observation scheme. We can always make these numbers equal to each other.

Note also that the other method did failed to find the phase function even at $\tau = 0.8$.

In Tables 1 and 2 the values

$$\Delta_i = \frac{g_i - g_i^*}{g_i^*}$$

Table 1: Identification results at $\tau = 0.8$

γ°	Iteration		
	0	1	2
0	1.55	-0.02	0.01
2	1.25	-0.03	0.01
6	0.96	0.05	0.01
10	0.80	0.04	-0.02
20	0.64	0.06	0.01
30	0.42	0.05	0.03
40	0.28	0.10	0.02
60	0.08	0.07	-0.03
80	-0.03	-0.05	0.00
100	-0.14	0.09	-0.01
140	-0.20	0.08	0.02
180	-0.05	0.12	0.03

Table 2: Identification results at $\tau = 1.5$

γ°	Iteration					
	0	1	2	3	4	5
0	1.55	0.21	0.10	0.03	0.05	-0.02
2	1.25	0.15	0.06	0.05	0.03	0.01
6	0.96	-0.10	0.06	-0.09	0.02	-0.03
10	0.80	-0.05	-0.12	-0.08	-0.04	-0.01
20	0.64	0.33	-0.07	0.10	-0.07	0.01
30	0.42	-0.23	0.14	0.05	0.10	-0.03
40	0.28	0.11	-0.02	0.09	0.08	-0.02
60	0.08	-0.08	0.11	-0.06	-0.12	0.00
80	-0.03	0.19	0.03	0.02	0.07	0.02
100	-0.14	-0.01	0.09	-0.04	-0.06	-0.01
140	-0.20	0.16	0.08	0.02	0.08	-0.02
180	-0.05	0.04	0.02	-0.03	0.11	0.03

of relative errors of the estimated components g_i of the phase function are presented. As one might expect, the method is more effective for the smaller values of the optical thickness of the layer. Two iterations were enough to find the desired values with error less then 3 % at $\tau = 0.8$. The same result for the problem with $\tau = 1.5$ required five iterations.

We should mark a specific feature of the estimation process. To obtain a rough estimate in the first iterations we can calculate the matrix coefficients with comparatively low accuracy. But as the iterations approach to the exact solution the accuracy of the calculation must improve. In the numerical experiment the relative error of the calculation did not exceed 1 %. We can see from the tables that the error of estimation (of about 2–3 %) has the same order as the error in the right-hand side of the system of equations, which is the error of the "measurement" (about 1 %). Hence, the error of identification in the considered cases is significantly caused by the error of the "measurement" data.

Bibliography

Albert, G. E. (1956). A general theory of stochastic estimates of the Neumann series for the solution of certain Fredholm integral equations and related series. In: *Symposium of Monte Carlo Methods.* Wiley, New York.

Anderson, . (1963). *Introduction to the many-dimensional statistical analysis.* Fizmatgiz, Moscow (in Russian).

Anikonov, D. S. (1985). Uniqueness of the determination of a coefficient of the transport equation for special type of source. *Sovet Mat. Dokl.* **32** (in Russian).

Antyufeev, V. S. (1977). On solving some stochtastic problems in atmosphere optics by the Monte Carlo method. *Izv. AN SSSR FAO* **13**, No. 5 (in Russian).

Antyufeev, V. S. (1979). Method of mathematical expectations in solving problems on radiation transfer having plane geometry. *J. Vich. Mat. i Mat. Fiz.* **19**, No. 6 (in Russian).

Antyufeev, V. S. (1980). On convergence of one iteration process. In: *Chisl. Met. Stat. Model. v Teorii Perenosa.* Nauka, Novosibirsk (in Russian).

Antyufeev, V. S. (1984). Calculation of the correlation characteristics of radiation field in the stochastic medium by the Monte Carlo method. *J. Vich. Mat. i Mat. Fiz.* **24**, No. 10 (in Russian).

Antyufeev, V. S. (1986). Solving integral equation of the 2-nd kind with stochastic kernel by the Monte Carlo method. *J. Vich. Mat. i Mat. Fiz.* **26**, No. 4 (in Russian).

Antyufeev, V. S. (1992). Direct and inverse problems of the radiation transfer in a plant canopy. *Rus. J. Num. Anal. and Math. Mod.* **7**, No. 3.

Antyufeev, V. S. (1993). On estimating the leaf-normal distribution for a plant canopy. *Remote Sens. Env.* **41**.

Antyufeev, V. S. (1994). Optimization in solving inverse problems of atmospheric optics by the Monte Carlo method. *Rus. J. Num. Anal. and Math. Mod.* **9**, No. 6.

Antyufeev, V. S. (1996). Solution of the generalized transport equation with highly peaked phase function by the Monte Carlo method. *Rus. J. Num. Anal. and Math. Mod.* **11**, No. 2.

Antyufeev, V. S., Avaste, O. A., *et. al.* (1979). Recovery of the altitudinal variations of the extinction aerosole coefficient using space optical measurements. In: *Proc. Study of Optical-Atmospheric Conditions from Board of the Orbital Station "Salyut-4"*. Tartu (in Russian).

Antyufeev, V. S., Avaste, O. A., *et. al.* (1981). Recovery of the altitudinal variations of the aerosole coefficient and water vapor using space measurements. In: *Proc. Study of Optical-Atmospheric Conditions from Board of the Orbital Station "Salyut-4"*. Tartu (in Russian).

Antyufeev, V. S., and Bondarenko, A. N. (1996). X-ray tomography in scattering media. *SIAM Journal of Applied Math.* **56**, No. 2.

Antyufeev, V. S., Ivanov, A. I., *et. al.* (1980). Determination of aerosol scattering indicatrices for a cloudless atmosphere in the spectral region 0.55–2.4 m. *Izv. Akad. Nauk SSSR, Ser. Fiz. Atmos. i Okeana* **16**, No. 2 (in Russian).

Antyufeev, V. S., Laktionov, A. G., *et. al.* (1992). Spectral characteristics of the ocean-atmosphere system. *Issled. Zemli iz Kosmosa* **3** (in Russian).

Antyufeev, V. S. and Marshak, A.L. (1988). Use of the Monte Carlo method for calculating of radiation transfer in plant canopy. In: *Proc. Int. Sem. on Rad. in Canopy.* Tartu.

Antyufeev, V. S. and Marshak, A. L. (1989). Monte Carlo method and transport equation in plant canopies. *Optika Atmosfery* **11** (in Russian).

Antyufeev, V. S. and Marshak, A. L. (1990a). Monte Carlo method and transport equation in plant canopies. *Remote Sens. Environ.* **31**.

Antyufeev, V. S. and Marshak, A. L. (1990b). Inverse problem for plant parameters estimating. *Remote Sens. Environ.* **33**.

Antyufeev, V. S. and Marshak, A. L. (1991). Monte Carlo model for transport equation in a plant canopy and its inversion. In: *Proc. 5th Int. Coll.-Phys. Measur. Sign. in Rem. Sen., Courchevel, France.*

Antyufeev, V. S. and Mikhailov, G. A. (1976). Recovery of the altitudinal variations of the scattering coefficient and calculation of correlation characteristics for the brightness of horizon by the Monte Carlo method. *Izv. Akad. Nauk SSSR, Ser. Fiz. Atmos. i Okeana* **12**, No. 5 (in Russian).

Antyufeev, V. S. and Nazaraliev, M. A. (1987). Solving some inverse problems in atmosphere optics by the Monte Carlo method. In: *Proc. Numerical Methods and Mathematical Modelling.* Moscow (in Russian).

Antyufeev, V. S. and Nazaraliev, M. A. (1988a). *Inverse Problems of Atmosphere Optics.* Computing Center of Siberian Branch of the USSR Academy of Sciences, Novosibirsk (in Russian).

Antyufeev, V. S. and Nazaraliev, M. A. (1988b). Solving of inverse problems in atmosphere optis by the Monte Carlo method. *Sov. J. Num. Anal. and Math. Mod.* **3**, No. 2.

Atroshenko, V. S., Glazova, K. S., *et. al.* (1962). Calculation of light intensity in the atmosphere at anisotropic scattering. In: *Proc. IFA AN SSSR. P. 2, Vol. 3* (in Russian).

Batles, H. P. (1980). *Inverse Scattering Problems in Optics.* Springer-Verlag, New-York.

Berger, M. J. (1955). Reflection and transmission of gamma radiation by barriers: semianalytic Monte Carlo calculaton by a collision-density method. *J. Res. Nat. Bur. Stand.* **55**.

Bondarenko, A. N. (1992). Structure of singularities of the fundamental solution of transport equation and inverse problems of particle scattering. *Dokl. AN SSSR* (in Russian).

Bunnik, N. J. J. (1978). The multispectral reflectance of shortwave radiation by agricultural crops in relation with their morphological and optical properties. *Mededelingen Landbouwhogeschool* **78-1**.

Calos, M. H. and Whitlock, P. A. (1986). *Monte Carlo Methods.* Wiley, New York.

Candelore, N. R. and Gast, R. C. (1964). *A Monte Carlo Program for Estimating Epithermal Capture Rates in Rod Arrays.* Report WAPD-TM-427, Bettis Atomic Power Laboratory, Los Alamos.

Case, K. M., Hoffman F., and Placzek, G. P. (1953). *Introduction to the Theory of Neutron Diffusion.* Los Alamos Scientific Laboratory, Los Alamos.

Case, K. M. and Zweifel, P. F. (1967). *Linear Transport Theory.* Addison–Wesley, Reading.

Gerstl, S. A. W. (1986). Off-nadir optical remote sensing from satellites for vegetation identification. In: *Proc. IGARSS'86.* Zurich.

Curtis, J. N. (1954). Monte Carlo methods for iteration of linear operators. *J. Math. Phys.* **32**.

Davison, B. (1958). *Neutron Transport Theory.* Clarendon Press, Oxford.

Deans, S. R. (1983). *The Radon Transform and Some of its Applications.* Wiley, New York.

Ermakov, S. M. (1975). *Monte Carlo Method and Related Problems.* Nauka, Moscow (in Russian).

Feygelson, E. M., Malkevich, M. S., *et. al.* (1962). Calculation of light intensity in the atmosphere at anisotropic scattering. *Proc. IFA AN SSSR. P. 1, Vol. 3* (in Russian).

Franklin, J. N. (1963). Deterministic simulation of random processes. *Math. Comput.* **17**.

Glazov, G. N. and Titiv, G. A. (1980). Equation for the moments of radiation intensity in broken clouds. *Izv. Vuzov, Radiofizika* **23**, No. 4 (in Russian).

Glushko, V. N., Ivanov, A. I., *et. al.* (1974). *Scattering of IR-Radiation in a Cloudless Atmosphere.* Nauka, Alma-Ata (in Russian).

Goel, N. S. (1988). Models of vegetation canopy reflectance and their use in estimation of biophysical parameters from reflectance data. *Remote Sens. Reviews* **22**.

GOST 25645.153-90 (1991). *Scattered Radiation of the Earth's Atmosphere. The Model of Spatial-Temporal Distribution.* Izd. Standartov, Moscow (in Russian).

Helgason, S. (1980). *The Radon Transform.* Birkhauser-Verlag, Basel.

Grunbaum, F. A., Herman, G. T., *et. al.* (1990). *Diffuse Tomography: A Refined Model in Mathematical Methods in Tomography.* Springer-Verlag, Oberwolfach.

Herman, G. T. (1979). *Image Reconstruction from Projections, Implementation and Application.* Springer-Verlag, Berlin – New York.

Herman, G. T., Natterer F. N., and Louis, A. K. (1990). *Mathematical Methods in Tomography.* Springer-Verlag, Oberwolfach.

Holton, J. H. (1970). A retrospective and prospective survey of the Monte Carlo methods. *SIAM Rev.* **12**.

Hotelling, H. (1933). Analysis of a complex of statistical variables in principal components. *J. Educ. Psych.* **24**.

Hummersley, J. M. and Handscomb, D. C. (1964). *Monte Carlo Methods.* Mehuen, London.

Kondrat'yev, K. Ya., Marchuk, G. I., *et. al.* (1977). *The Field of Atmospheric Radiation.* Izd. LGU, Leningrad (in Russian).

Kozlov, V. L., Malkevich, S. M., and Turchin, V. F. (1970). Application of methods of mathematical statistics to the solution of ill-posed problems. *Usp. Mat. Nauk* **3**, No. 2 (in Russian).

Kuusk, A. (1985). The hot-spot effect of a uniform vegetative cover. *Remote Sens. Environ.* **3**.

Marchuk, G. I. (1964). On posing some inverse problems. *Dokl. Akad. Nauk SSSR* **156**, No. 3 (in Russian).

Marchuk, G. I. and Mikhailov, G. A. (1967). On solving problems of atmospheric optics by the Monte Carlo method. *Izv. Akad. Nauk SSSR, Ser. Fiz. Atmos. i Okeana* **3**, No. 3 (in Russian).

Marchuk, G. I., *et. al.* (1980). *The Monte Carlo Method in Atmospheric Optics.* Springer-Verlag, New York.

Marshak, A. L. (1990). Effect of the hot-spot on the transport equation in plant canopies. *J. Quant. Spectrosc. Radiat. Transfer* **42**.

Mikhailov, G. A. (1967). Calculation of derivatives of functionals of the transport equation solution over the system parameters by the Monte Carlo method. *J. Vich. Mat. i Mat. Fiz.* **7**, No. 4 (in Russian).

Mikhailov, G. A. (1974). *Some Problems of the Theory of Monte Carlo Methods.* Nauka, Novosibirsk (in Russian).

Mikhailov, G. A. (1992). *Optimization of the Monte Carlo Weight Methods.* Springer-Verlag, New York.

Myneny, R. B., Ross, J. K., and Asrar, G. (1989). A review on the theory of photon transport in leaf canopies. *Agric. Forest. Meteor.* **45**.

Natterer, F. N. (1986). *The Mathematics of Computerized Tomography.* B. G. Teubner, Stuttgart.

Nilson, T. and Kuusk, A. (1989). A reflectance model for the homogeneous plant canopy and its inversion. *Remote Sens. Environ.* **27**.

Pomraning, G. C. (1973). *The Equations of Radiation.* Pergamon, Oxford.

Radiation Commission IAMAP, WMO.WCP-112 (1986). *A Preliminary Cloudless Standard Atmosphere for Radiation Computation.* Boulder, Colorado.

Romanov, V. G. (1987). *Inverse Problems of Mathematical Physics.* VNU Science, Utrecht.

Ross, J. K. (1981). *The Radiation Regime and Architecture of the Plant Stands.* Dr. W. Junk Publ., The Hague.

Ross, J. K. and Marshak, A. L. (1988). Calculation of canopy idirectional reflectance using the Monte Carlo method. *Remote Sens. Environ.* **24**.

Ross, J. K. and Marshak, A. L. (1989). The influence of leaf orientation and specular component of leaf reflectance on the canopy bidirectional reflectance. *Remote Sens. Environ.* **27**.

Sheddon, J. N. (1972). *The Use of Integral Transforms.* McGraw-Hill, New York.

Shultis, J. K. and Myneny, R. B. (1988). Radiative transfer in vegetation canopies with anisotropic scattering. *J. Quant. Spectrosc. Radiat. Transfer* **39**.

Smith, K. T. and Keinert, F. (1985). Mathematical foundations of computed tomography. *Appl. Optics* **24**.

Smith, K. T., Solomon, D. S., and Wagner, S. L. (1977). Practical and mathematical aspects of the problem of reconstructing objects from radiographs. *Bull. Amer. Math. Soc.* **83**.

Sobol, I. M. (1973). *Numerical Monte Carlo Methods.* Nauka, Moscow (in Russian).

Spanier, J. and Gelbard, E. M. (1969). *Monte Carlo Principles and Neutron Transport.* Addison-Wesley, Reading.

Tantashev, M. V. (1982). Solution to the transport equation for random-inhomogeneous medium by Monte Carlo. *J. Vich. Mat. i Mat. Fiz.* **22**, No. 4 (in Russian).

Turchin, V. F. (1967). Solving the Fredholm equation of the 1-st kind for a statistical ensemble of smooth functions. *J. Vich. Mat. i Mat. Fiz.* **7**, No. 6 (in Russian).

Turchin, V. F. and Nozik, V. Z. (1969). Statistical regularization for solving ill-posed problems. *Izv. AN SSSR, FAO* **5**, No. 1 (in Russian).

Vladimirov, V. S. (1981). *Equations of Mathematical Physics.* Nauka, Moscow (in Russian).

Bei Fragen zur Produktsicherheit wenden Sie sich bitte an:
If you have any questions regarding product safety,
please contact:

Walter de Gruyter GmbH
Genthiner Straße 13
10785 Berlin
productsafety@degruyterbrill.com